MODELS AND MODELING IN THE SCIENCES

Biologists, climate scientists, and economists all rely on models to move their work forward. In this book, Stephen M. Downes explores the use of models in these and other fields to introduce readers to the various philosophical issues that arise in scientific modeling. Readers learn that paying attention to models plays a crucial role in appraising scientific work.

This book first presents a wide range of models from a number of different scientific disciplines. After assembling some illustrative examples, Downes demonstrates how models shed light on many perennial issues in philosophy of science and in philosophy in general. Reviewing the range of views on how models represent their targets introduces readers to the key issues in debates on representation, not only in science but in the arts as well. Also, standard epistemological questions are cast in new and interesting ways when readers confront the question, "What makes for a good (or bad) model?"

All examples from the sciences and positions in the philosophy of science are presented in an accessible manner. The book is suitable for undergraduates with minimal experience in philosophy and an introductory undergraduate experience in science.

Key features:
- The book serves as a highly accessible philosophical introduction to models and modeling in the sciences, presenting all philosophical and scientific issues in a nontechnical manner.
- Students and other readers learn to practice philosophy of science by starting with clear examples taken directly from the sciences.
- While not comprehensive, this book introduces the reader to a wide range of views on key issues in the philosophy of science.

Stephen M. Downes is Professor of Philosophy at the University of Utah, USA. He has published articles on the philosophy of biology, the biology of human behavior, and scientific models. He is co-editor, with Edouard Machery, of *Arguing about Human Nature* (Routledge, 2013).

MODELS AND MODELING IN THE SCIENCES

A Philosophical Introduction

Stephen M. Downes

NEW YORK AND LONDON

First published 2021
by Routledge
52 Vanderbilt Avenue, New York, NY 10017

and by Routledge
2 Park Square, Milton Park, Abingdon, Oxon, OX14 4RN

Routledge is an imprint of the Taylor & Francis Group, an informa business

© 2021 Taylor & Francis

The right of Stephen M. Downes to be identified as author
of this work has been asserted by him in accordance with
sections 77 and 78 of the Copyright, Designs and Patents
Act 1988.

All rights reserved. No part of this book may be reprinted
or reproduced or utilised in any form or by any electronic,
mechanical, or other means, now known or hereafter
invented, including photocopying and recording, or in
any information storage or retrieval system, without
permission in writing from the publishers.

Trademark notice: Product or corporate names may be
trademarks or registered trademarks, and are used only for
identification and explanation without intent to infringe.

Library of Congress Cataloging-in-Publication Data
A catalog record for this title has been requested

ISBN: 978-1-138-12222-2 (hbk)
ISBN: 978-1-138-12223-9 (pbk)
ISBN: 978-1-315-64745-6 (ebk)

Typeset in Bembo
by codeMantra

Printed in the United Kingdom
by Henry Ling Limited

CONTENTS

List of Figures		*vii*
Preface		*ix*
1	Introduction	1
2	Models in the Sciences	6
3	Characterizing and Classifying Models	33
4	Models and Representation	52
5	What Makes for a Good (or Bad) Model?	69
6	Conclusion: Pluralism about Models, Modeling, and Model Evaluation	84
References		*91*
Index		*99*

FIGURES

2.1	Watson and Crick's model of DNA	7
2.2	DNA. Redrawn from Watson & Crick (1953b), 965	8
2.3	DNA–RNA process. © udaix / Shutterstock	9
2.4	Exponential growth curve for a theoretical population	9
2.5	Exponential growth of global population from 1300 to 2000 (x-axis), in billions of people (y-axis)	10
2.6	A logistic growth curve in a theoretical population. Adapted from OpenStax College, Biology CC BY 4.0	10
2.7	Two other applications of logistic curves: (a) haploid and (b) diploid population growth in yeast. Redrawn from Otto and Day (2007)	11
2.8	Oscillations in population size. From Weisberg (2013)	13
2.9	Graph of Hare and Lynx skin sales to the Hudson Bay Company across time, appearing to show a predator prey curve. Data from Odum (1953). Graph redrawn based on Lamiot CC BY 4.0	13
2.10	Neutral evolution. © Jonsta247 CC BY 4.0	14
2.11	An example of an Edgeworth box. © SilverStar CC BY 3.0	16
2.12	A classic supply/demand curve, as used in economics. © Feco CC BY-SA 3.0	17

viii Figures

2.13	Fleeming Jenkin's supply and demand curve experiments. Redrawn from Morgan (2012)	18
2.14	Output from Schelling's segregation model. Redrawn from Weisberg (2013)	19
2.15	Position and velocity functions in a mass and spring system. Redrawn from Giere (1988)	21
2.16	(a) Heliocentric model. Redrawn from De revolutionibus orbium coelestium. (b) Newtonian model. © Shutterstock	22
2.17	(a) Rutherford model of the atom. (Electrons in light grey, nucleus in dark grey.) Redrawn based on Ben Steele CC BY-SA 3.0. (b) Bohr model of the atom. Redrawn based on Cdang CC BY-SA 3.0	23
2.18	Basic greenhouse effect model picture. Robert A. Rohde CC BY-SA 3.0	25
2.19	San Francisco Bay model. US Arms Corps of Engineers	26
2.20	A capacity model for attention (Kahneman 1973, 10)	27
2.21	Lorenz's psychohydraulic model of motivation. Redrawn (based on Slater 2004, 59) after Lorenz, K.Z., 1950 Symp. Soc. Exp. Biol. 4, 221–68	28
2.22	A schematic model outlining some of the events that lead to drinking following extracellular fluid loss. Redrawn from *The Hypothalamus*, edited by L. Martini, M. Motta, and F. Fraschini, Academics Press, 1970. (Sourced via Cotman and McGaugh 1980, 576)	30
2.23	McKinley and Johnson's model of thirst (McKinley and Johnson 2004, 3)	31
2.24	A mouse. © Igor Stramyk / Shutterstock	31
3.1	Mathematical models in science journals	38
3.2	Math models vs. organismal models	39
3.3	Giere's model of relationships among sets of statements, models, and real systems. Adapted from Giere (1988), 83	46
3.4	Godfrey-Smith's proposed modifications to Giere's diagram. Redrawn from Godfrey-Smith (2006), 733 [Modified from Giere (1988)]	47

PREFACE

Given the enormous amount of literature on models and modeling in science, writing an introduction to this area is a daunting task. What I have tried to do here is introduce some of the key topics of discussion about models and modeling in the philosophy of science. The material covered in the chapters and the bibliography is nowhere near comprehensive. Rather, what we have here will give students and interested professionals a route into the vast literature on models and modeling in philosophy of science.

The idea for this introductory book on models in science was that of Andy Beck, Routledge Philosophy Editor, and I thank him, his Editorial Assistant, Marc Stratton and Jeanine Furino in production, for steering this project through to completion. Thanks are also due to Rebekah Cummings and Allyson Mower at the University of Utah Marriott Library for bibliographic help and help with finding figures and securing permissions for them. I also thank all the students in one of my philosophy of science classes on models for helping me dry run this material and making numerous helpful suggestions. Thank you: Melissa Alvalos, Jake Borcher, Chelsea Bush, Tiffany Campbell, Morgan Carlson, Jennifer Edwards, Dominique Faalogo, Philip Handley, Nick Hebdon, Walter Joseph, Matt Miller, Kaitlin Pettit, Jared Poole, Clint Simkins, Lacey Slizeski, Kristin Taylor, and Jeremy Wiedemeier. A big thanks to all of you who have given your time to

x Preface

discuss issues surrounding modeling with me over the years including: Nancy Cartwright, Jordi Cat, Melinda Fagan, Arthur Fine, Patrick Forber, Roman Frigg, Peter Godfrey-Smith, Jim Griesemer, Kristen Hawkes, Phillip Kitcher, Elisabeth Lloyd, John Matthewson, James Nguyen, Cailin O'Connor, Jay Odenbaugh, Wendy Parker, Angela Potochnik, Anya Plutynski, Kyle Stanford, Kim Sterelny, Mauricio Suarez, Martin Thompson-Jones, Jim Weatherall, Michael Weisberg, Bill Wimsatt, Eric Winsberg, and Rasmus Winther.

I would like to thank referees Michael Weisberg and Eric Winsberg as well as several anonymous referees for all of their helpful comments on various stages of this manuscript.

Finally, a big thanks goes to the Philosophy Department at the University of Utah both for providing the financial support that helped me to complete this project and for providing a wonderfully supporting work environment. While working on this project, I have been the recipient of a Charles Monson Esteemed Faculty Award, a sabbatical leave, and an administrative leave.

1

INTRODUCTION

Scientists use models to further scientific knowledge. If you do a search for papers involving models in a scientific database, you will return hundreds of thousands of hits. Scientists in almost every discipline develop models, test models, and attempt to confirm models. There are advanced textbooks on how to model phenomena in biostatistics, climate science, economics, evolutionary biology, molecular biology, and neuroscience, to name just a few.

Philosophers of science aim to understand science. We have epistemological interests in science: we want to understand what makes scientific knowledge special. Here the focus can be on the way in which hypotheses are tested or theories are supported. We also have an interest in understanding and articulating the worldview that scientists present: what is the underlying structure of things and processes we are familiar with via our senses? Almost all aspects of science proceed or are assisted by modeling. As a result, modeling and models are a focus of philosophers of science. In this book, we introduce a selection of the philosophical issues that arise from a focus on model-based science.

Philosophers of science first introduced models as a counter to what they considered an overly sentence-based understanding of science. Logical positivists and logical empiricists presented and defended a highly influential approach to science that characterized

2 Introduction

scientific theories as collections of sentences representing laws and observation sentences. This general approach provided the framework for tackling philosophical issues in the sciences such as the confirmation of scientific theories and the way in which scientific theories explain. Tools from deductive logic, and to a lesser degree probability theory, were used to characterize confirmation relations between theories and observations or between explananda (sentences describing phenomena to be explained) and explanans (sentences explaining these phenomena). Models were introduced as a corrective, or, for some, a supplementation, to this overall approach to the philosophy of science. Philosophers argued that scientific inferences would be better understood by examining relations between models and their targets and, further, that scientific theories would be better characterized in terms of models rather than as collections of sentences.

In the mid- to late 1980s, some philosophers of science turned their attention to scientific practice. In contrast to the philosophers characterized above, these philosophers paid careful attention to the way in which scientists deployed models in their theoretical or experimental work (see e.g. N. Cartwright 1983) as well as in the teaching of science through textbooks (see e.g. R. Giere 1988). While these philosophers continued to invoke models in challenging the sentence-based approach to science and its attendant problematic, their work was increasingly driven by attention to the details of model development and deployment by scientists. This approach revealed a myriad of philosophical issues in science and pointed to new ways of approaching and resolving these issues. Influenced by these more recent developments in the philosophy of science, this book starts out with a collection of sample models. In Chapter 2, we present a range of models from various sciences. These examples illustrate both different approaches to modeling and also different applications of related modeling approaches in different fields. For example, mathematical modeling is revealed to have many different aims depending upon the field in which the mathematical models are deployed. In many areas of physics, mathematical models are taken to represent a system in the world, whereas in population genetics, evolutionary biology, and ecology, mathematical models often are designed to bring to light new hypotheses for testing. In this chapter, we describe models from each of these areas of science. Scientists do not limit the scope of modeling practice to just mathematical modeling. Molecular biologists' modeling is diagram-based, and increasingly 3D visualization techniques are used in presenting molecular biology models. Cognitive scientists also rely on diagrammatic models. Each of these approaches is also presented in Chapter 2. Presenting a nice range of varied examples of models and modeling practice sets up discussions in subsequent chapters. The discussion in Chapters 3–5 appeals to our sample models set out here.

In Chapter 3, we review various attempts by philosophers and scientists to classify models. Many classification schemes have been proposed for models each of which has advantages and disadvantages. Some propose only two model types, material and formal models, while others propose a multiplicity of model types. Providing a classification of model types is part of the project of answering the question, "What are models?" Before turning to this question, we briefly look at relations between models and theories. Much philosophical interest in models was first driven by the search for alternate accounts of scientific theories. As we saw above, logical empiricist, and other related, philosophers of science defended a view of theories as sets of sentences. Some critics of this view of theories proposed that theories were more appropriately viewed as collections of models. We briefly review this situation and show how the idea that theories are collections of models shaped philosophical discussion about models in science. While the idea that theories are collections of models has gained some traction, there are problems for this view if proponents think that models can only be the kinds of things that comprise theories. In this chapter, we look at the challenge to this view from philosophers who examine models developed by scientists in nontheoretical contexts. Finally, we look at some answers to the question, "What are models?" For example, some say that models are abstract objects and some say that models are fictions. A few prominent answers to this question are assessed here. We then go on to present answers to this question that are more practice-based. The practice-based approach leads to answers to the question, "What are models?" in terms of the uses scientists put models to. We reveal that this approach leads many philosophers to conclude that there are multiple model types and multiple answers to the question, "What are models?" We also consider various opposing views to this liberal or pluralist position.

4 Introduction

There is a pervasive view that scientists produce models to represent parts of the world and that, as a result, models are to be assessed in terms of their representational capacities. In Chapter 4, we lay out the representational view of models and then introduce and critically appraise some alternate accounts of the model representation relation. We consider some of the leading accounts of representation such as similarity accounts and agent-based accounts. We also consider accounts that combine features of these two approaches. The merits of various accounts of scientific representation are also discussed against the background of discussions of views of representation in general. Here we introduce and appraise the idea that there should be no special account of representation for models or other representational vehicles in science. We then discuss nonrepresentationalist approaches to characterizing models and modeling in philosophy of science. This is the idea that some models, or some aspects of models, do not do their epistemic work via representation. One contrast that is brought out between the nonrepresentationalists and representationalists is that the former agree that some models represent but some do not, while the latter tend to think that all models represent. We also consider the ways in which discussions of models as representations proceed somewhat differently depending on whether they are generated in the context of the discussion of specific examples of models taken from scientific practice or whether they are pursued from a more abstract perspective.

Given that models are proposed for many purposes in the sciences, it is likely that there are varying methods for evaluating models and varying grounds for appraising them. One group of scientists can prioritize models that best match relevant empirical data, while others prioritize models with very little apparent fit to such data. Models are valued for their predictive power, their explanatory power, their productivity in hypothesis generation, their troubleshooting capabilities, and in some cases, their likeness to a real-world system. In Chapter 5, we examine a selection of this wide range of approaches to model appraisal. We assess appraising models via their accuracy, the extent to which they are confirmed, and their explanatory power. We also look at the way in which model desiderata can trade off; for example, if we want to maximize the generality of our model, we cannot also maximize its precision. Finally, we examine the idea of robustness

Introduction **5**

for groups of models. We reveal in this chapter that a focus on models adds many dimensions to classic epistemological discussions about science. For example, if we start from the assumption that all scientific theories have the same structure and can be characterized in the same way, discussions of confirmation are constrained to the confirmation of theories, so characterized, by the empirical evidence. Once we acknowledge that there are many different approaches to modeling resulting in many different types of models, it is not as easy to see how one notion of confirmation will be apt in all situations in the sciences. Similar points apply for explanation. In this chapter, we uncover this expansion of the epistemology of science brought about by a focus on models.

We conclude with a short chapter on pluralism about models and modeling. Our discussions in the previous chapters lead to the conclusion that there are many types of models that serve many epistemological roles in the sciences. As we will see from discussions in Chapters 3–5, there are philosophers who defend monistic approaches to models; for example, some defend the view that all models are abstract representations of the phenomena of interest. Our discussion, driven by consideration of many models from many different sciences, does not support such a monistic approach. We conclude that scientists use many different model systems for many different ends.

2

MODELS IN THE SCIENCES

In 1953, Francis Crick and James Watson published a paper outlining a model of the structure of DNA (1953). One version of the model they described was constructed from sticks and beads (somewhat reminiscent of a Tinkertoy structure) forming a double helical structure (Figure 2.1). The model portrayed the "building blocks" of DNA and provided insight into the way in which DNA strands pulled apart so that RNA molecules could be built by matching the pattern of nucleotides in the relevant segment of DNA. Much work in molecular biology is still guided by the Watson and Crick model of DNA. Scientists no longer use beads and sticks; instead, they use visual representations based on the structures captured in the stick and bead model (Figure 2.2). Models of gene expression have the same components as Watson and Crick's model (Figure 2.3). Modeling genes as strands of nucleotides – adenine, guanine, cytosine, and thymine (A, G, C, T) in DNA and A, G, C and uracil (U) in RNA – is now standard practice throughout biology.

There are two models of population growth used by biologists and social scientists. In both models, individuals reproduce and die but the models differ in their assumptions about available resources. In an exponential growth model individuals' access to resources is constant whereas in a logistic growth model access to resources decreases as the population increases. Equation (2.1) is a differential equation in continuous time for exponential growth:

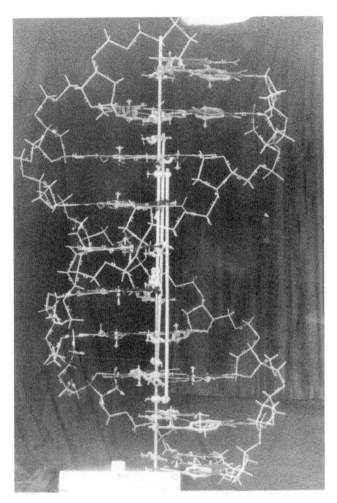

FIGURE 2.1 Watson and Crick's model of DNA.

$$\frac{dn}{dt} = r\, n\,(t) \tag{2.1}$$

Figure 2.4 illustrates a typical shape for an exponential growth curve. For this model of growth, we assume that each individual produces two offspring. Alarmingly, the best model for human population growth data is exponential growth as illustrated in Figure 2.5.

8 Models in the Sciences

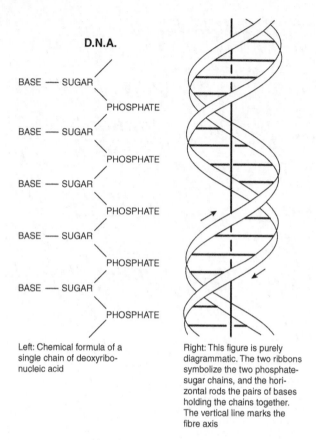

FIGURE 2.2 DNA. Redrawn from Watson & Crick (1953b), 965.

The key to the logistic growth model is the interaction between population size and available resources. Equation (2.2) is a differential equation in continuous time for logistic growth and includes the additional term K, which denotes the carrying capacity of the population:

$$\frac{dn}{dt} = r\, n(t) \left(1 - \frac{n(t)}{K}\right) \tag{2.2}$$

The shape of a logistic growth curve is illustrated in Figure 2.6. The carrying capacity of a population is the size at which population growth flattens out or individuals merely replace one another. There

Models in the Sciences **9**

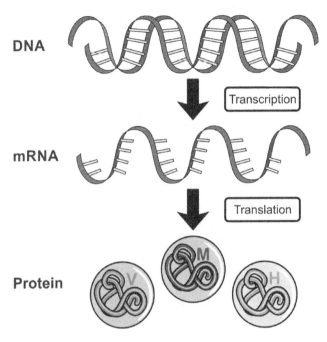

FIGURE 2.3 DNA–RNA process. © udaix / Shutterstock.

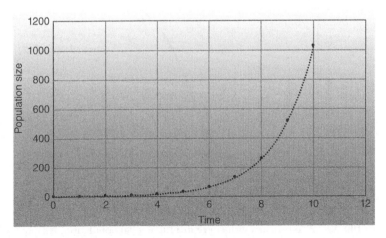

FIGURE 2.4 Exponential growth curve for a theoretical population.

10 Models in the Sciences

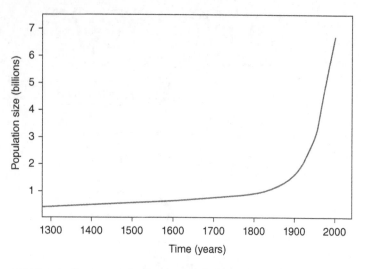

FIGURE 2.5 Exponential growth of global population from 1300 to 2000 (*x*-axis), in billions of people (*y*-axis).

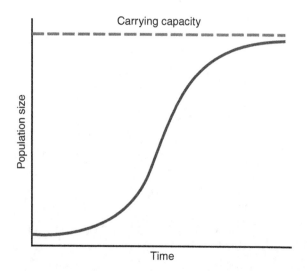

FIGURE 2.6 A logistic growth curve in a theoretical population. Adapted from OpenStax College, Biology CC BY 4.0.

are lots of examples of population growth in biology that are best modelled as logistic growth. Figure 2.7 shows two applications of the logistic growth model to biological data. These growth models can be applied in a wide range of contexts. For example, economists can

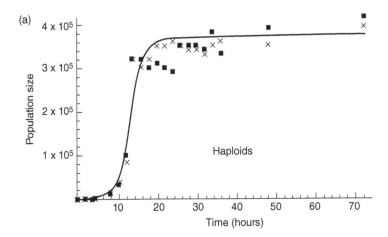

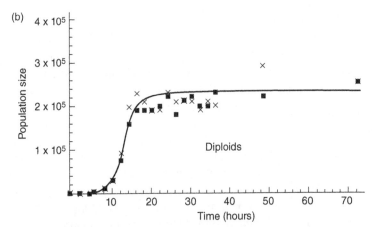

Example of logistic population growth in yeast. Population sizes are plotted over time based on data from (a) haploids and (b) diploids. Crosses and squares represent data from two different replicates. The solid curves are plots of the population size over time based on equations and the estimated parameter values.

FIGURE 2.7 Two other applications of logistic curves: (a) haploid and (b) diploid population growth in yeast. Redrawn from Otto and Day (2007).

12 Models in the Sciences

model growth in sales of any commodity as either logistic or exponential. Also, the logistic model includes an exponential component. One way of thinking about a logistic curve is as a flattening exponential curve.

While population growth models have wide application, biologists understand that the models assume an unchanging environment and assume no interaction between the individuals in the population being modelled and other populations of individuals. Competition models are a class of models that can account for relations between populations (Otto and Day 2007). A well-known competition model is the Lotka-Volterra predator prey model. The Lotka-Volterra model builds upon the logistic growth model adding the impacts of one species upon another on each species' population growth. Weisberg (2013) presents a version of the Lotka-Volterra model in terms of the following coupled differential equations:

$$\frac{dV}{dt} = rV - (aV)P \tag{2.3}$$

$$\frac{dP}{dt} = b(aV)P - mP$$

Here r is the growth rate of the prey population, and m is the death rate of the predator population; V is the size of the prey population, and P is the size of the predator population. Figure 2.8 illustrates the coupled oscillations in population size given by the model, and Figure 2.9 is a graphical presentation of animal skin sales data, strongly suggesting interacting populations of hares and lynxes that looks somewhat similar to the graph produced by the model (Michael Weisberg 2013, 93). The model is able to predict the outcome of various impacts on both populations such as that of large-scale biocide. For example, the model predicts that overfishing will lead to a disproportionate increase in prey fish populations over predator fish populations.

Many models in evolutionary biology focus on variation in the structure of organisms' DNA rather than on variation in morphological or behavioral traits. These models incorporate components of molecular models, descendants of Watson and Crick's model, along

Models in the Sciences **13**

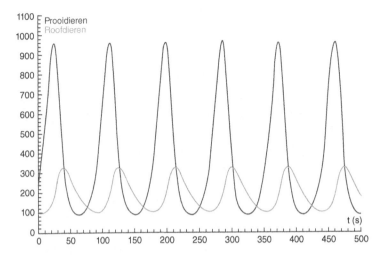

FIGURE 2.8 Oscillations in population size. From Weisberg (2013).

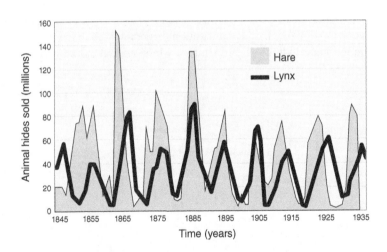

FIGURE 2.9 Graph of Hare and Lynx skin sales to the Hudson Bay Company across time, appearing to show a predator prey curve. Data from Odum (1953). Graph redrawn based on Lamiot CC BY 4.0.

14 Models in the Sciences

with components of models of changes in populations such as those reviewed above. Change between generations of organisms or differences between different species of organisms can be assessed in terms of numbers of nucleotide substitutions. Nucleotide substitutions can be deleterious, in the case in which a coding nucleotide triplet is changed into a non-coding triplet, but they can also be silent. There is redundancy in DNA (and RNA); for example, the DNA nucleotide triplet TTC is transcribed as AAG in RNA and codes for the amino acid lysine. TTT is transcribed as AAA in RNA but also codes for lysine. Hence, the substitution of T for C has no impact on the organism. Figure 2.10 illustrates some substitution options for TTC. Neutral models of evolution specify rates of accumulation of variation in nucleotides at various loci in genomes occurring without selection and with no impact on fitness (Futuyma 2006; Lynch, and Walsh 1998). Neutral models set a baseline for mutation rate and if biologists discover divergences from this mutation rate they can propose different sources of the relevant variation in the population (Wimsatt 2007, 100). Divergences in mutation rate can signal that different types of selection have occurred.

To model neutral evolution we assign a probability μ to the event that a mutation (nucleotide substitution) becomes fixed in the population by chance, which gives the rate at which neutral mutations arise.

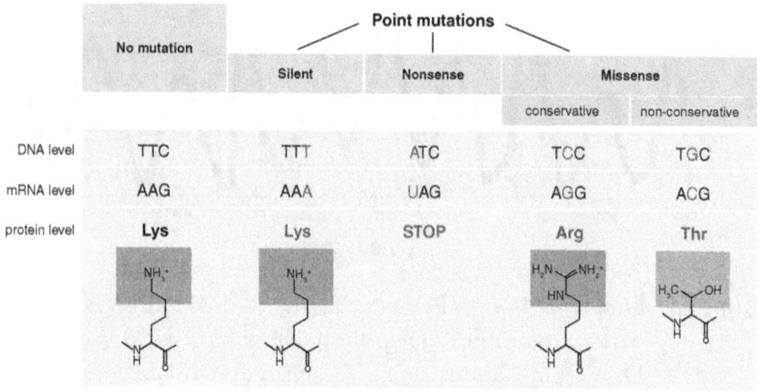

	No mutation	Point mutations			
		Silent	Nonsense	Missense	
				conservative	non-conservative
DNA level	TTC	TTT	ATC	TCC	TGC
mRNA level	AAG	AAA	UAG	AGG	ACG
protein level	**Lys**	Lys	STOP	Arg	Thr

FIGURE 2.10 Neutral evolution. © Jonsta247 CC BY 4.0.

The rate of substitution per generation, K, in a diploid population of size N is expressed in equation (2.4):

$$K = 2N\mu \times \frac{1}{2N} = \mu \qquad (2.4)$$

$1/2N$ expresses the probability that a mutation occurs, which is its relative frequency in the population. From this, we can give the fraction of mutations that have become fixed, x, after t generations as:

$$x = \mu t \qquad (2.5)$$

Evolutionary biologists measure mutation rates in populations of organisms, and if they find $K > \mu$, it is evidence of positive selection.

Economists rely on models as much as biologists. We talk about models from economics in our daily discourse, often without being aware that we are invoking a model. For example, when we speak of supply and demand we are invoking a model or rather one simple version of a whole host of models (see Morgan 2012 for details). Economists are confronted with analogous phenomena to those faced by biologists. Each has to account for relations between populations of agents (organisms), each other, and available resources. The Edgeworth box (or Edgeworth–Bowley box) is a way of modeling the choice space of two agents with two sets of fixed available resources. In Figure 2.11, our agents are Octavio and Abby and the resources available between them are 10 liters of water and 20 hamburgers (Edgeworth's agents were Robinson Crusoe and Man Friday). Every possible division of the available good can be represented as a point in the box. This model has been developed and built upon by many economists in something like the way that the logistic growth model has been built upon by evolutionary biologists (Morgan 2012) notes that Pareto was an early developer of Edgeworth Box models).

Models of supply and demand are also nested models based on Mangoldt's models of Adam Smith's account of the relation between the price of a commodity and the supply and demand for that commodity (Morgan 2012). The relation between supply and demand is given by paired equations of quantity of goods and price

16 Models in the Sciences

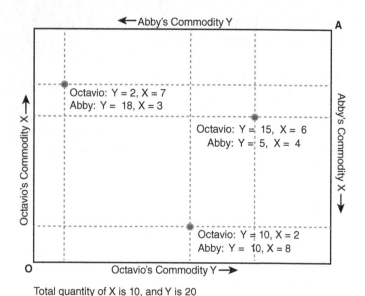

Total quantity of X is 10, and Y is 20

FIGURE 2.11 An example of an Edgeworth box. © SilverStar CC BY 3.0.

of goods. A simple version of the model is given by the following equations:

$$Q_d = a - bP \qquad (2.6)$$

$$Q_s = e + fP$$

Along with the equilibrium point:

$$Q_s = Q_d$$

Q_d is the demand, and Q_s is the supply. The demand curve has a negative slope, to account for the idea that there will be less demand for a $100 veggie burger than a free one.

Figure 2.12 illustrates a typical supply–demand relation.

This simple model can be expanded in many ways to accommodate shifting supply, changing demand structure and so on as illustrated in

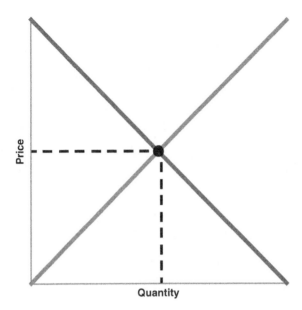

FIGURE 2.12 A classic supply/demand curve, as used in economics. © Feco CC BY-SA 3.0.

Figure 2.13, a presentation of Fleeming Jenkin's work on the supply and demand model (Morgan 2012).

John Maynard Smith's hawk/dove game theoretic model of animal behavior further illustrates the overlap in modeling techniques and approaches between biology and economics. Maynard Smith (J. Maynard Smith and Price 1973; John Maynard Smith and Parker 1976) used a game theoretic model to answer questions about the evolution of strategies for animal confrontation over resources. What Maynard Smith's models revealed was that there were different kinds of equilibrium points but most involved populations engaging in multiple strategies. An evolutionarily stable state is an equilibrium at which an invading mutant cannot gain a foothold in a population, such as when an allele is at fixation in a population (Otto and Day 2007). There are thousands of game theoretic models in use in the social sciences, and a subgroup of these are evolutionary game theoretic models like Maynard Smith's. Evolutionary game theoretic models

18 Models in the Sciences

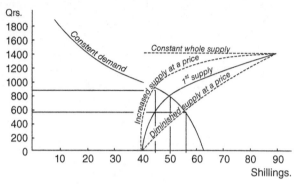

2.13a: Showing Changes in the Supply Curve.

If the supply curve rise to the upper dotted line the market price will fall to 45, and 900 quarters of wheat will change hands, instead of 800.

If the supply curve fall as shown by the lower dotted line the market price will rise to 55, but only 600 quarters will be sold.

The whole supply, the price at which all would be sold or none sold, all bought or none bought, may all remain unaltered, as well as the demand curve. In practice some or all of these elements would generally vary when the supply curve varies.

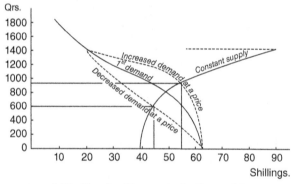

2.13b: Showing Changes in the Demand Curve.

If the demand curve rise, as in the upper dotted line, the market price will be 55, and the quantity sold be 900 quarters.

If the demand curve fall, as in the lower dotted line, the price will fall to 45, and the quantity sold to 600 quarters.

As in 2.13a, the whole supply, the price at which all would be sold or none sold, all bought or none bought, is left unaltered, as well as the supply curve.

In practice some or all of these elements would generally vary when the demand curve varies.

FIGURE 2.13 Fleeming Jenkin's supply and demand curve experiments Redrawn from Morgan (2012).

assess how well a strategy does over many iterations, where doing well is cashed out via a notion of fitness or reproductive success. These models are not one-shot games, such as one shot Prisoner's Dilemmas in Political Science and Philosophy; rather, computer simulations of multiple iterations of the game situation are produced. If we have a population of Hawks and Doves whose competition for a resource V can incur a cost C, a pay-off matrix for Hawks v Doves is as follows:

	Hawk	Dove
Hawk	V/2-C/2	V
Dove	0	V/2

If the cost of losing is higher than the value of winning and a value is set for fitness, there is an equilibrium percentage of Hawks in the population of C/V.

Another way of modeling behavior in populations is via agents on a spatial grid. In such grids, rules govern agents' choice of neighbors. Schelling's model of segregation is such a model (discussed in both Grüne-Yanoff 2013; Weisberg 2013). Schelling's original model of segregation used dimes and nickels to represent individuals and a chessboard represented location. Now, such models are all computational. We can construct such models for ourselves using a multiprogramming modeling environment such as NetLogo. Where the Hawk Dove game reaches an equilibrium of strategies in a population, the Schelling model reaches an equilibrium of distribution of agents in space relative to one another. In Schelling's model, each agent prefers that at least 30% of its neighbors are of the same type; A's prefer that at least 30% of their neighbors are A's. In a spatial grid, this boils down

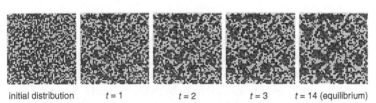

initial distribution $t = 1$ $t = 2$ $t = 3$ $t = 14$ (equilibrium)

An example of Schelling's model of segregation on a 51 x 51 grid with 2000 agents. Each agent prefers 30% of its Moore neighbors to be the same shape and color. The initial distribution of agents was random, and the model equilibrated after fourteen time steps.

FIGURE 2.14 Output from Schelling's segregation model. Redrawn from Weisberg (2013).

20 Models in the Sciences

to preferring three of your neighbors being of the same type and being indifferent if three to eight are of the same type. Figure 2.14 illustrates an initial condition and a subsequent equilibrium distribution for blue and red agents. Schelling proposed his model as a model of racial segregation in cities without any presumption of racism on the part of agents (Weisberg 2013, 13).

We now turn to some examples from physics. Mechanics can be construed as involving a connected collection of models or family of models (Giere 1988). A simple model in mechanics is the linear oscillator or harmonic oscillator. The model is characterized by a version of Newton's Second Law:

$$F = -kx \tag{2.7}$$

The force on a particle is proportional to its negative displacement from a resting position. This model can be illustrated by Hooke's Law for a spring, which says that the force of a spring is proportional to the amount it is stretched (Giere 1988, 68). Figure 2.15 shows a diagram of a mass and spring system and the shape of the position and velocity functions for the system. In the mass and spring system, k in equation (2.7) is a measure of the stiffness of the spring. The spring in the model is taken to have no mass, the displacement of the mass is taken to be linear, neither the mass nor the spring is subject to frictional forces, and the spring is attached to a completely rigid object at both ends. Equations for a real spring would have to include some notion of damping, which is included in the equation for a damped harmonic oscillator. A damped oscillator model still contains many idealizations and does not exactly model the behavior of a mass and spring system.

One main difference between Newton's model of the solar system and that of, one of his predecessors, Copernicus (see Figure 2.16a and b. for diagrams of both systems) is that in Newton's model, the positions of the planets are given in part by iterated applications of the equation of motion for a two-body system. This equation is derived from the same equations of motion for masses that the harmonic oscillator equation is derived from. Newton's model proved to have a much greater predictive capacity than those of his predecessors, because observed perturbations in a planet's orbit could be accounted for by adding another planet to the system, via the equations of motion, to account for the perturbations. Newton was also able to derive

Models in the Sciences   21

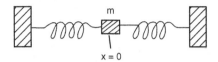

Figure 2.15a. The mass and spring system used to illustrate Hooke's law and simple harmonic motion.

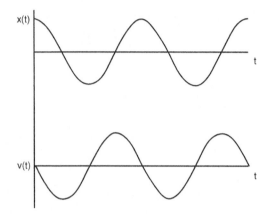

Figure 2.15b. Position and velocity as functions of time for the mass and spring system of figure 2.15a, with initial conditions $x\,(t = 0) = A$ and $v\,(t = 0) = 0$.

FIGURE 2.15 Position and velocity functions in a mass and spring system. Redrawn from Giere (1988).

Kepler's proposed elliptical motion of the planets from his equations of motion. In contrast, for Copernicus all observed planetary motion was to be accounted for in terms of circular motion only. This model had less predictive or explanatory flexibility than Newton's model.

The Bohr model of the atom is analogous to a solar system model. In the early 20th century, Rutherford performed experiments on gold foil that led physicists to believe that atoms consisted of a dense positively charged central region with small negatively charged particles, proportionately far away from the central region. Rutherford's model of the atom (see Figure 2.17a) presents the distance between the nucleus and the electrons and has the electrons orbiting the nucleus but does not model the structure and organization of the electrons in their orbits. Bohr's model (see Figure 2.17b) introduced components of quantum mechanics to model stable orbits for electrons and provided

22 Models in the Sciences

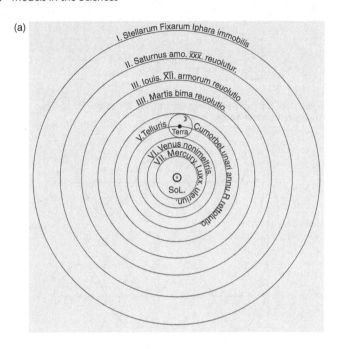

FIGURE 2.16 (a) Heliocentric model. Redrawn from *De revolutionibus orbium coelestium*. (b) Newtonian model. © Shutterstock.

Models in the Sciences   23

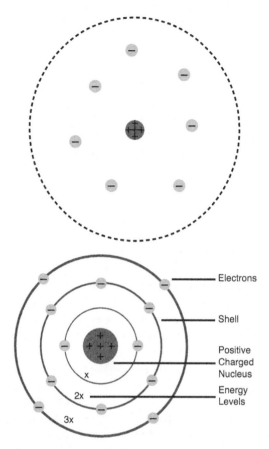

FIGURE 2.17   (a) Rutherford model of the atom. (Electrons in light grey, nucleus in dark grey.) Redrawn based on Ben Steele CC BY-SA 3.0. (b) Bohr model of the atom. Redrawn based on Cdang CC BY-SA 3.0.

an account of electron movements between stable orbiting positions or shells. A drawback of the Rutherford model was that electrons would lose energy while orbiting and fall into the center of the atom. Now neither model is considered the correct model of atomic structure but the Bohr model is still used in physics and chemistry textbooks to introduce students to concepts in elementary quantum mechanics.

24 Models in the Sciences

Our sample models from mechanics all involve one or more equations. They can be viewed as expressing many alternate possible relations between states or many alternate dynamical patterns to a system. The equations introduced in statistical models also model a pattern but are introduced along with a null model. A simplified example illustrates this point. Say we are interested in the variance in a quantity in individuals in a population and we think that some variable is responsible for the variance in that quantity. For example, the height of maize plants varies and can vary as a result of watering regimens. Human height also varies and some analyses propose that genes are the main determinant of this variance. Equation (2.8) expresses a model for this kind of situation:

$$Y = A + \varepsilon \tag{2.8}$$

Here Y is the response variable, height in maize plants, A is the factor producing the variation in height, the watering regimen, and ε the residual response in the variable. The simplest analysis of variance (ANOVA) models, called one-factor or one-way ANOVA models, introduce one factor (Doncaster and Davey 2007). The further component to an ANOVA test is the testing of the model against a null model, such as equation (2.9):

$$Y = \varepsilon \tag{2.9}$$

In ANOVA, an F statistic (or F ratio or F test) is used to calculate the probability P that the observed response (variation in plant height) could be produced by the null model. Our factor, the watering regimen, is deemed significant if $P < \alpha$, where α is usually set at 0.05. The right-hand side of equation (2.8) can be expanded to add other candidate sources of variance in the response variable Y. For example, plant height could also vary as a result of seed density or proximity of plants to one another. Expanding the right-hand side of equation (2.8) to reflect this situation gives to a two-factor ANOVA model, resulting in increased complexity in the testing process. Testing ANOVA models is generally done with a computational statistics package (such as SPSS).

Multifactor ANOVA models are known to be complex but climate models are notoriously complex and difficult to present in simple form. Climate models likely receive more press than any other scientific models.[1] Climate models inform (or should inform) debate about policy for mitigating and adapting to the impacts of climate change and extreme weather events. Here, rather than presenting an example of one climate model in detail, we look at some of what goes into producing a climate model and some of the background assumptions involved in building climate models (McGuffie and Henderson-Sellers, 2005).[2] A sample of the range of phenomena climate scientists model includes ocean currents, ocean temperature change, global temperature change, sea-level change, and polar icecap melt. One collection of climate models is based on the notion of greenhousing, which in turn is based upon the difference between global absorption and radiation of heat. Figure 2.18 presents a simplification of the greenhousing process. The earth's chief source of warmth is the sun. Solar radiation is absorbed into the earth and some of the energy from this radiation is reflected. The reflected energy is of much longer wavelength than the energy from the sun and does not easily pass out of the earth's atmosphere back into space. The atmosphere contains water and carbon dioxide along with other gases and particulates. If the contents of the atmosphere contain higher levels of gases and particulates, less energy is able to radiate out into space.

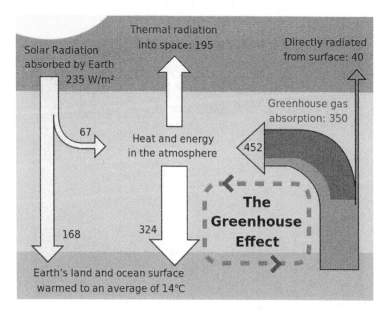

FIGURE 2.18 Basic greenhouse effect model picture. Robert A. Rohde CC BY-SA 3.0.

26 Models in the Sciences

Climate scientists produce models for predicting the rate of increase of global warming and such models vary levels of particulates and relative concentrations of gases in the atmosphere. For example, use of fossil fuels increases the concentration of carbon dioxide in the atmosphere. Climate models involve computer simulations rather than analytic solutions to the large number of differential equations for the various processes involved in climate systems (see Winsberg 2010, 2018). Climate models can produce predictions of global temperature increase by a specific date and model relations between the increase in the products of fossil fuel burning in the atmosphere and temperature increases.

Other modeling techniques are also used to investigate issues arising from humans and our environment. In response to a proposal to dam the San Francisco Bay to capture fresh water from the Sacramento and San Joaquin rivers, the Army Corps of Engineers constructed a scale model of a damn of the San Francisco Bay (Weisberg 2013). The model is housed in a large warehouse covering around 1.5 acres (Figure 2.19). The model uses hydraulic pumps to simulate the flow of both fresh water and salt water into the Bay. The Corps Engineers can model "tides, currents, and the gradient where fresh and salt water

FIGURE 2.19 San Francisco Bay model. US Arms Corps of Engineers.

meet" (Weisberg 2013, 7). Like Watson and Crick's model, the Bay model is a concrete structure but unlike the DNA model, the Bay model includes many moving parts. In the Bay model, measurements are taken on the movement of water and the impact of water movement on the salinity of the Bay model among other things.

We now turn to a selection of models from cognitive science and behavioral neuroscience. Daniel Kahneman's (1973) capacity model of attention (Figure 2.20) presents attention as a fixed resource or capacity that is required for the execution of other mental tasks. Activities we engage in such as visual recognition aided by memory or solving simple arithmetic problems require attention. Some recent work on visual attention (see e.g. Strayer and Drews 2008) is guided by a version of the Kahneman model.

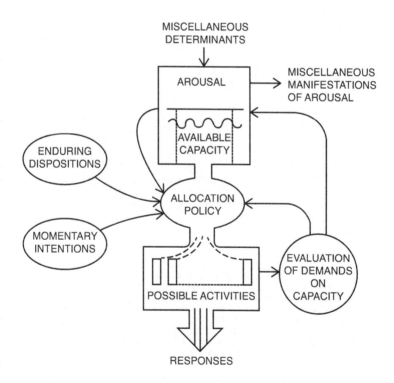

FIGURE 2.20 A capacity model for attention (Kahneman 1973, 10).

28 Models in the Sciences

Kahneman's capacity model of attention is similar in structure to more explicitly hydraulic models,[3] such as Konrad Lorenz's "psychohydraulic" model of motivation, illustrated in Figure 2.21 (Slater 2004, 59). In Kahneman's model, there is a reserve of attention that flows to or can be directed to specific tasks. Lorenz explicitly models the energy reservoir in his model as a water supply. Water, or energy, accumulates

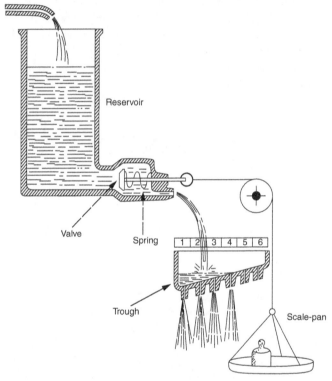

Lorenz's psychohydraulic model of motivation. Action specific energy is represented by water, which accumulates progressively in a reservoir when the behaviour concerned is not being expressed. The behaviour pattern occurs when the water passes out of the reservoir into the trough beneath. Higher threshold aspects of the behaviour (numbered 4, 5 and 6) are only shown when a lot of water is passing into the trough. The valve is so arranged that it is opened by the combined effect of the water in the reservoir (action-specific energy) and of weights on a scale pan, which represent the adequacy of external stimulation.

FIGURE 2.21 Lorenz's psychohydraulic model of motivation. Redrawn (based on Slater 2004, 59) after Lorenz, K.Z., 1950 Symp. Soc. Exp. Biol. 4, 221–68.

Models in the Sciences **29**

when the animal is not engaged in the relevant behavior, such as seeking food. On this model a satiated animal has low motivation to seek for food and a hungry animal has accumulated a reserve of energy sufficient to drive food-seeking behavior. Also, Lorenz's model predicts that animals can be prompted to behave in the relevant way by very mild stimuli if their energy reserve is full. Both Kahneman's and Lorenz's models guide experimental work. On one interpretation both models are models of mechanisms but work in philosophy on mechanisms indicates care is needed when characterizing models of mechanisms. (Craver and Tabery 2019 provide a summary of philosophical work on mechanisms.) Kahneman's and Lorenz's models are both likely best understood as mechanism sketches (Darden 2002) as they provide some idea of the possible organizational structure of the mechanisms of attention and motivation but do not model any of the detail of the relevant underlying processes.

Models of thirst in behavioral neuroscience help to illustrate the distinction philosophers of mechanism make between mechanism schemas (see e.g. Darden 2002) and mechanical models (Glennan 2005). Cotman and McGaugh's (1980) model of thirst (see Figure 2.22) provides a good example of Darden's mechanism schema. Cotman and McGaugh's model can be filled in with detailed descriptions of the processes and relations between them to produce a more complete mechanical model (see also Craver and Tabery 2019).

There are many similar diagrams to Figure 2.22, depicting schematic models of thirst, in many specific organisms, for example, in humans. These models present dehydration in terms of blood volume decrease and blood osmolality increase. No detail of these processes is represented in schematic models (see e.g. Figure 2.22), and neither are any of the correlated hormonal signalling and neural processes. In contrast, McKinley and Johnson's (2004) model of thirst is a much more detailed mechanical model (see Figure 2.23). Their model presents both the neural and hormonal inputs to the mammalian brain that produce thirst as well as the integration of these signals and their translation into motor output. While this model has much more detail than the schematic models of thirst, it is not a complete description of all the relevant detail of all the neural processes involved in mammalian thirst.

Our final example of a model is mdx mouse (see Figure 2.24), which is a model organism[4] for Duchenne muscular dystrophy (DMD) (see e.g. McGreevy et al. 2015). Like the DNA model and the

30 Models in the Sciences

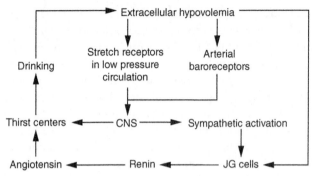

A schematic model outlining some of the events that lead to drinking following extracellular fluid loss. A fall in blood volume is detected by stretch receptors, arterial baroreceptors, and by the juxtaglomerular cells (JG cells) of the kidney. Activation of the stretch receptors and baroreceptors stimulates the CNS which, in turn, activates the thirst centres in the hypothalamus and the sympathetic division of the ANS. This, in turn, adds additional stimulation to the JG cells which release renin into the bloodstream. Renin is converted to angiotensin II which also stimulates thirst centres to induce drinking. Drinking behaviour then reduces the extracellular volume deficit.

FIGURE 2.22 A schematic model outlining some of the events that lead to drinking following extracellular fluid loss. Redrawn from *The Hypothalamus*, edited by L. Martini, M. Motta, and F. Fraschini, Academics Press, 1970. (Sourced via Cotman and McGaugh 1980, 576).

San Francisco Bay model, mdx mouse is a concrete object but is also a living organism. DMD is a progressive muscle-wasting disorder. DMD sufferers lack dystrophin, which is a crucial muscle protein. DMD sufferers have a mutation (or mutations) in a gene also called DMD. There is currently no cure for the disease in humans. Mdx mice have a point mutation that does not allow the full expression of dystrophin. Young mdx mice start out life with normal muscle development but after two weeks or so they experience severe muscle degeneration. The deterioration of mdx mouse muscles goes through phases, including some periods of time during which muscles develop normally. At around 15 months mdx mice experience muscle wasting and heart failure. There are a large number of different strains of mdx mice, each of which expresses variants of muscle wasting often coupled with other debilitating symptoms. A strain of mouse is taken to be a better model of human DMD if the trajectory of symptom development more closely matches that of human DMD sufferers.

Models in the Sciences **31**

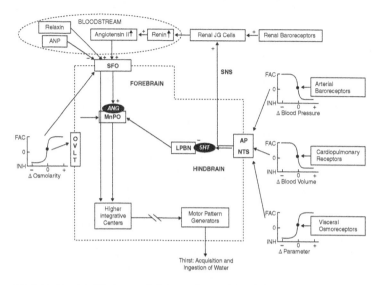

FIGURE 2.23 McKinley and Johnson's model of thirst (McKinley and Johnson 2004, 3).

FIGURE 2.24 A mouse. © Igor Stramyk / Shutterstock.

These models provide the raw material for the philosophical discussions in the chapters that follow. In the next chapter, we turn to the discussion about alternate classifications of models and the question, "What are models?"

32 Models in the Sciences

Notes

1 In 2020 most of our attention and the attention of the press is focused on epidemiological models of the spread of COVID 19.
2 This approach is consistent with that of philosophers of science who focus on climate models (W. Parker 2009; Lloyd 2010; Steele and Werndl 2016; Winsberg 2010, 2018).
3 Economists also construct hydraulic models, such as the Phillips-Newlyn hydraulic model of money flow (Mary S Morgan 2012, 35).
4 See Leonelli and Ankeny (2013) and Fagan (2016) for discussion of the use of organisms as models in biological research.

Further Reading

Philosophers of science take their examples of models from various sources including scientific textbooks (see e.g. Giere 1988) and scientific research papers (see e.g. Morgan 2012; Morrison 2015). However, there is also a vast literature on how to construct models in various fields. The following is a brief selection of such texts, which can give a good idea of alternate model construction processes and alternate modeling resources in some of the sciences:

Ciobanu, G., & Rozenberg, G. (2004). *Modeling in molecular biology*. Berlin: Springer.

Cobelli, C., & Ewart, C. (2008). *Introduction to modeling in physiology and medicine*. Elsevier.

Doncaster, C. P., & Davey, A. J. H. (2007). *Analysis of variance and covariance: how to choose and construct models for the life sciences*. Cambridge: Cambridge University Press.

Ingalls, B. P. (2013). *Mathematical modeling in systems biology: an introduction*. Cambridge, MA: M.I.T. Press.

Jonoska, N. editor, Saito, M. editor, & Ebooks Corporation. (2014). *Discrete and topological models in molecular biology*. Heidelberg: Springer.

McGuffie, K. (2005). *A climate modeling primer* (3rd ed.). Hoboken, NJ: John Wiley.

Neelin, D. J. (2011). *Climate change and climate modeling*. Cambridge: Cambridge University Press.

Otto, S. P., & Day, T. (2007). *A biologist's guide to mathematical modeling in ecology and evolution*. Princeton, NJ: Princeton University Press.

Numerous online documents are also available detailing model construction techniques and resources in varying fields including physics, which is not represented in our list of texts here.

3

CHARACTERIZING AND CLASSIFYING MODELS

3.1 Introduction

The scope of application of the term "model" is very broad, leading to Nelson Goodman's description of the wide range of models: "Few terms are used in popular and scientific discourse more promiscuously than 'model'. A model is something to be admired or emulated, a pattern, a case in point, a type, a prototype, a specimen, a mock-up, a mathematical description – almost anything from a naked blonde to a quadratic equation – and may bear to what it models almost any relation of symbolization" (1976, 171). He went on to say: "scientists and philosophers [...] have been forced to fret at some length about the nature and function of *models*" (1976, 171; italics in original). The scope of application of the term "scientific model," while still broad, is not as broad as this. Scientific models are characterized in part not only by their structure (see e.g. Weisberg 2013) but also by what scientists do with them. Models are also characterized by the modeling practice in which they are developed. Maynard Smith (along with Price and others) simultaneously developed a new modeling practice for biology when proposing his game of theoretic models of the evolution of animal behavior. Characterizing models in terms of their relations to their objects can lead to a very broad scope notion of models. This approach can also narrow the scope of "model." For example, structural realists (see e.g. Ladyman 1998, 2007) characterize

34 Characterizing and Classifying Models

models partly in terms of the structures that they pick out and are isomorphic to (or partially isomorphic to). This narrows the range of items appropriately thought of as models. Attending to modelers and their practice can make the object of study clearer, but this approach can also lead to issues of scope. For example, there is a strong tradition in the life sciences of treating organisms as models. Our mdx mouse is an example of such a model. If we include such objects among the scope of "scientific model," we encounter difficulties in coming up with a unified account of models. Rejecting such examples without providing good grounds can also lead to problems.

In this chapter, we consider some of the ways in which philosophers and scientists have confronted these problems. In Section 3.2, we look at various attempts to classify models into types. Next, in Section 3.3, we look at various accounts of the relation between models and theories, each of which commits their proponents to different views of what models are or how they can be classified. Finally, in Section 3.4, we turn to various answers to the question, "What are models?"

3.2 Types of Models

Philosophers and scientists have proposed many classifications of models into types. This project could prove very helpful in sorting out the commonalities and differences between all of our sample models in Chapter 2 and, by extension, all of the thousands of models presented in scientific publications. Let's first consider Arturo Rosenblueth and Norbert Weiner's (1945) proposal that we analyze models via a distinction between material and mathematical models. A material model "is the representation of a complex system by a system which is assumed simpler" (1945, 317). Material models also are assumed to have "some properties similar to those selected for study in the original complex system" (1945, 317).[1] A formal model "is a symbolic assertion in logical terms of an idealized relatively simple situation sharing the structural properties of the original factual system" (1945, 317). Can we use this distinction to sort our models? We can sort many of our models using this distinction. For example, DNA, the Bay Bridge, dmx mouse, and Lorenz's motivation model are all material models on this account and Edgeworth Box, Schelling's model, Lotka-Volterra, and Hawk/Dove

Characterizing and Classifying Models **35**

are all formal. However, some of our models do not fall neatly on one side or the other of this distinction. For example, while Copernicus' solar system model fits the definition of a material model, Newton's solar system model has both formal and material components. Also, harmonic oscillators appear to be both material and formal models.[2] This issue partly arises from separating a model from its description (model description) and we will pursue this in a little more detail below. The issue also arises from a lack of specificity in the definitions. Let's focus first on the definition of a formal model. It is easy to take Rosenblueth and Weiner's formal models to be more or less synonymous with mathematical models, but their definition could include models in any number of formal systems, including computer algorithms, deductive logic, natural language syntax, probability, and a wide range of mathematical models. Deductive logic is not a good scientific modeling resource (see Downes 1992; Thomson-Jones 2010) whereas calculus is a very good modeling resource. More can be said about distinguishing between the range of formal models captured in the definition. Further, distinctions can be made between material models. For example, mice are animals, material objects, whereas Lorenz's motivation model is something like an unactualized mechanism schema. Let's look at some other model classification schemes and assess whether they address some of these issues.

Scientists reflect on the nature of models and their classification. For example, Richard Lewontin (1963) discusses the nature of models and their role in biology and emphasizes the importance of this discussion for both practicing scientists and philosophers of science.[3] Douglas Futuyma (2006) proposes a classification for models in biology in his textbook on evolutionary biology. Futuyma introduces five types of models: verbal models, physical models, graphical models, mathematical models, and computer models. Verbal models (cf. Winther 2006) are sets of statements characterizing a state of affairs. Tversky's model of attention is the closest of our models to a verbal model. The statements characterizing the model invoke no mathematics or computation and so the model can be thought of as verbal. A physical model is a physical system used to find something out about a target system. Our DNA model is a physical model, and possibly mdx mouse counts as a physical model on Futuyma's account. A graphical model is a graph plotted from data. Futuyma says that mathematical models "are

36 Characterizing and Classifying Models

often the most powerful, and are favored in all the sciences, including evolutionary biology" (1998, 229). We have many examples of mathematical models in Chapter 2 including the Lotka-Volterra model, the Logistic Growth model, and the Harmonic Oscillator. Finally, Futuyma says that computer models are used when "mathematical models are too complex to solve analytically" (1998, 229). Computer models simulate the systems we are studying. Arguably our Global Climate models are computer models in Futuyma's sense of the term (Winsberg 2010, 2018).

Futuyma's classification of models is a hierarchical one. Rather than proposing that there are distinct types of models, he introduces mathematical models as an improvement on graphical models, which in turn are an improvement on physical models and so on. Many scientists and some philosophers share this idea that there are not distinct types of models but alternate models that can be improved upon by being presented mathematically. For example, Daniela Bailer-Jones (2009) characterizes the move from "mechanical models" to mathematical models as a progressive move in the history of science. If (even only in principle) all alternate models can be rendered mathematically, then perhaps the only type of models we should countenance are mathematical models.[4]

Saying that mathematical models are the only type of models that should be acknowledged can be understood in a number of ways. On one interpretation the claim says what the best models are. There are many scientists (and some philosophers) who take this approach. The approach often comes with a view about how science should be properly pursued and what some philosophers call a demarcation criterion (see Callender and Cohen (2006) for more on demarcation in this context). The idea here is that science is distinguished from other practices by its use of mathematical models. Many resist this position, pointing out that much scientific knowledge has been achieved without the appeal to mathematical models. Philip Kitcher (see e.g. 1993) and Peter Godfrey-Smith (2006) both point out that Darwin's signature work appealed to no mathematical models. A more specific version of this point to this context can be made by pointing to the scientific achievements made with nonmathematical models. Steven French and James Ladyman (1999) respond to this line of argumentation by "biting the bullet" with respect to demarcation, committing themselves to the view that mathematical models are the only kinds of models that should be recognized and that science is only properly conducted via the use of mathematical models. Bas Van Fraassen's empiricist structuralism is aligned with Ladyman and French on this issue when he says that "The structuralism in "empiricists structuralism" refers solely to the thesis that all scientific representation is at heart mathematical" (2008, 239). These lines of debate are also found within sciences. For example, systems biologists present themselves as bringing superior tools, mathematical models, to bear in various areas of biology and some proposing that their work should supplant non-mathematical modeling approaches in those fields (see Fagan 2013; MacLeod 2018).

There are two weaker versions of the above claim about mathematical models. One is that mathematical models are the predominant type of models in the sciences and the other, connected to the first, is that wherever possible, a scientist should strive to produce a mathematical model. Let's look at each briefly. Mathematical models are found in work across the sciences and social sciences. For example, Sarah Otto and Troy Day (2007) reveal that in three important biology journals, *American Naturalist*, *Ecology*, and *Evolution*, 99% of articles made use of mathematical models in a broad sense of the term, that is they appealed to some form of mathematical analysis. The same survey reveals that 43% of the articles in these journals used mathematical models to obtain results; 100% of the articles published in *Ecology* and *Evolution* in the year sampled used either mathematical models to obtain results or some form of mathematical analysis. (Figure 3.1 illustrates the increase over time in numbers of mathematical models appealed to in papers in the biological sciences, and Figure 3.2 contextualizes this number by comparison with the number of papers appealing to model organisms over the same time period.) If we look to textbooks on modeling in various fields, these books predominantly deal with mathematical modeling. Modeling textbooks in fields as varied as climate science, molecular biology, physiology, and ecology and evolution, are all guides for constructing mathematical models. A large proportion of models in the sciences are mathematical models, but our nonmathematical models in Chapter 2 were also drawn from the sciences. So, while mathematical models may predominate, there are other types of models used by scientists.

38 Characterizing and Classifying Models

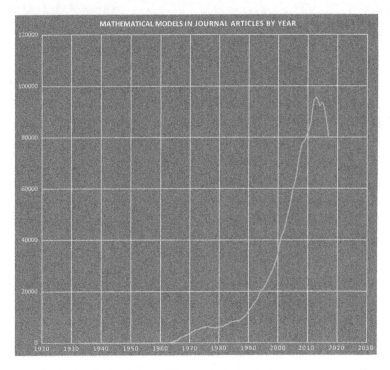

FIGURE 3.1 Mathematical models in science journals.

Further, the idea that scientists should strive to produce a mathematical model is not applicable in some contexts of inquiry. Some mechanical models (Machamer, Darden, and Craver 2000; Craver and Tabery 2019), for example, are not improved upon by rendering them in mathematical form and likewise model organisms. Also, some phenomena that have been modelled mathematically have also been modelled via agent-based computational models with equally revealing and helpful results (see e.g. Weisberg and Reisman's (2008) agent-based model of predator–prey relations). As a result, there still appears to be a need for a richer classification of models.

Michael Weisberg (2013) answers this need by proposing a classification of models into three types: concrete, mathematical, and computational. Weisberg's classification is not hierarchical like Futuyma's; rather, Weisberg bases his classification on distinctions between model

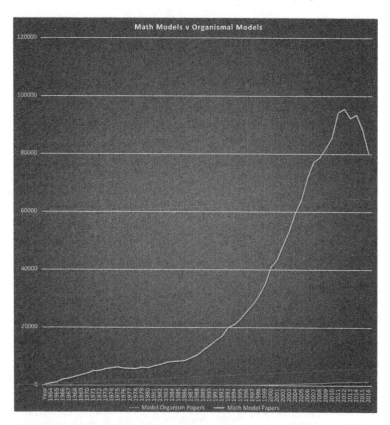

FIGURE 3.2 Math models vs. organismal models.

structures and properties. Let's flesh out Weisberg's classification a little more and examine it.

For Weisberg, models are interpreted structures and each model type has a different type of structure. Concrete models are real physical objects that stand in representation relations with parts of the world. Mathematical models contain mathematical structures that represent states and relations between states or state transitions (cf. Lloyd 1988). Computational models contain an algorithm (or algorithms), which is a set of instructions for carrying out a procedure (Weisberg 2013, 7). Weisberg uses three models (all introduced in our Chapter 2) to illustrate each of the model types: the San Francisco Bay model is a concrete

40 Characterizing and Classifying Models

model; the Lotka-Volterra model is a mathematical model; and the Schelling model is a computational model. Several more of our models fit straightforwardly into Weisberg's classification scheme. For example, the DNA model is a concrete model and the logistic growth model and harmonic oscillator are both mathematical models. Weisberg also argues that model organisms, such as mdx mouse, are concrete models. While many models can be assigned a model type in Weisberg's classification scheme without issue, there are problems confronting Weisberg's framework. Let's first focus on computational models.

Weisberg's characterization of computational models picks out a different group of models than Futuyma's computational models. For Futuyma (and lots of scientists), computational models are models that are used when the relevant mathematical equations do not have clear available solutions. For Weisberg, computer models contain algorithms. This means that the class of computer models is larger on Weisberg's account, than on Futuyma's account. The Schelling model, some evolutionary game theory models (such as our Hawks and Doves model), and agent-based models all contain algorithms, but none make use of computation specifically due to the lack of a clear solution to particular equations. Whereas the computer models used by climate scientists contain algorithms, which are deployed to aid approximate solutions to the relevant equations (Winsberg 2010, 2018). Weisberg acknowledges that computer models, on his account, "are just a special case of mathematical models" (2013, 19), because computer operations can be and are characterized mathematically. Despite this, he sustains his three-way distinction, rather than proposing that we settle for two model types: mathematical and concrete. Weisberg defends this move by emphasizing his focus on "epistemic" questions about models, rather than either purely descriptive or ontological questions. Weisberg's epistemic focus leads him to propose three kinds of models "to account for modeling as it is practiced in contemporary science" (2013, 20). This is in contrast to a purely descriptive approach that would countenance every model type scientists talk about in their daily practice or an ontological approach that answers the question of what models are fundamentally and how many types of those things there are. We will pursue the ontological question further in Section 3.4. Here, we consider another possible kind of model to add to Weisberg's classification scheme: statistical models.

Characterizing and Classifying Models **41**

Like computational models, statistical models are a special case of mathematical models, but some have argued that they should be distinguished from mathematical models on similar grounds to those Weisberg gives for distinguishing computer models from mathematical models. Elliott Sober (2008) proposes that we distinguish statistical models from "scientific models."[5] Statistical models "contain adjustable parameters; the statement that X and Y are related linearly is a model, while the statement that y = 3 + 4x is not" (2008, 79). Our ANOVA model from Chapter 2 is a statistical model as it presents a relation between plant height and watering regimen. Giere et al. (2006) says that statistical models are models of populations and that these models are of proportions, distributions, and correlations between values of variables in populations. Again, this would make our ANOVA model a statistical model as it presents a correlation between the values of the variables of height and watering regimen for plants. Statistical models clearly make use of mathematical tools but should they be considered a separate category of models? Cailin O'Connor and James Weatherall (2016) generalize this type of worry about Weisberg's classification of models into three types. They emphasize that there are many different types of mathematical models that are as distinct from one another as mathematical and computational models are on Weisberg's account and ask whether the line Weisberg draws between computational and mathematical models is perhaps an arbitrary one.

Weisberg anticipates a version of these kinds of objections arguing that his three-way distinction is up to the task of accounting for modeling as it is practiced. However, there are a range of alternate ways of categorizing models presented by philosophers and scientists aiming to account for modeling as it is practiced. Here is a sample of the array of model types presented by scientists and philosophers:

> probing models, phenomenological models, computational models, developmental models, explanatory models, impoverished models, testing models, idealized models, theoretical models, scale models, heuristic models, caricature models, didactic models, fantasy models, toy models, imaginary models, mathematical models, substitute models, iconic models, formal models, analogue models and instrumental models.
>
> *(Frigg and Hartmann 2018)*

42 Characterizing and Classifying Models

Roman Frigg and Stephan Hartmann's list of model types does not include some of the model types we have already discussed here and so we could add concrete models, mechanism models, material models, probability models, statistical models, and doubtless more. This proliferation of model types may reflect different approaches to classifying models. For example, Margaret Morrison and Mary Morgan (see e.g. Morgan and Morrison 1999; Morgan 2012; Morrison 2015) generate their classifications of models from the alternate ways that models function in scientific practice. Weisberg could argue that if you classify models in terms of their different structures, as he does, you would not end up with this vast array of model types. The problem here is that we have seen that even by restricting ourselves to structure alone, we can make a reasonable case for expanding (or contracting) Weisberg's three-way distinction.

Perhaps a shift of emphasis would help here. Rather than looking at an array of models presented in the sciences and trying to classify them into types, some philosophers propose that we answer the question "what are models?" first. Before we turn to this question, we take a look at the proposed relations between models and theories. Much work on models in philosophy of science arose out of criticisms of accounts of theories given by philosophers of science.

3.3 Models and Theories

Philosophers' interest in models was first driven by the search for alternate accounts of scientific theories (Suppe 1977a). Logical empiricist philosophers of science defended a view of theories as sets of sentences (see e.g. Hempel 1966 for an introduction to this approach), which Frederick Suppe (1977a) referred to as the "received view" of scientific theories. Much logical empiricist philosophy of science involved rationally reconstructing scientific practice as logical inferences on the sentences that were said to constitute theories (Hempel 1966). For example, on this view, theories involved laws and laws were characterized as universal generalizations. Confirmation of laws, understood as universal generalizations, was characterized using either deductive or inductive logic. While there were some notable exceptions, for example Mary Hesse (1966), philosophers of science in the 1960s and 1970s did not emphasize the role of models in scientific practice and instead

Characterizing and Classifying Models **43**

worked out the implications of the received view. Some critics of the received view of theories proposed that models, rather than sentences, should be the main focus of philosophers of science interested in explicating theories (see Suppe 1977a; and other authors in 1977b; as well as Van Fraassen 1980). Also, most accounts of models were closely tied to the idea of a model as the sequence that satisfies a set of sentences (Morrison and Morgan 1999b, 3). The sentences of the theory are true in the model. This approach derives from shifting focus from the sentences thought to characterize scientific theories to the models that satisfied these sentences (see e.g. Van Fraassen 1980). This approach to characterizing scientific theories is known as the semantic view of scientific theories (Suppe 1977a).

Ron Giere (1988) construes theories in terms of models, but his approach differs with those of previous semanticists in a number of ways. First, influenced by Nancy Cartwright (1983) and Ian Hacking (1983), among others, he turns his attention to scientific practice. The models Giere focuses on are those used by scientists in their daily practice and Giere pays particular attention to models as they are presented in science textbooks. Second, Giere rejects the idea that theories can be presented axiomatically. Previous semanticists, while rejecting the idea that scientific theories are collections of sentences, held onto the idea that the sentences involved in expressing scientific theories could be presented adequately via axiomatizations preserving a parallel between logic and science. Giere's idea is that theories are collections of models and that the relevant models have a much looser connection to the sentences invoked in a theory, than models do to an axiomatic system.[6] Giere's answer to the question "what is the structure of scientific theories?" is scientific theories are collections of models.[7] Paul Teller characterizes Giere's contribution here as the proposal "Theories are loose collections of models, organized inexactly in terms of the laws, methods, and other aspects which one uses to characterize models. One can also appeal here to the analogy of thinking of theories as 'model-building too kits'" (Teller 2001, 396).[8] On this account, articulating what models are now becomes part of the job of understanding scientific theories. Giere (along with Teller and others) effectively shifts the conversation in philosophy of science from what theories are (and how they can be true or false) to what models are and their role in scientific practice.

44 Characterizing and Classifying Models

Before we turn to the question of what models are, we should briefly reflect on the extent to which Giere's account of the relation between models and theories can help us address some of the issues that arose in the previous section. There we discussed various ways of classifying models. Is Giere's approach helpful for the project of classifying models discussed above and how many of the models introduced in Chapter 2 are best understood as constituents of theories in the relevant areas of scientific research? Alternately, must the models that make up theories be of a specific type, for example, mathematical as opposed to concrete?

First, are all our sample models best understood via their relation to theories? It appears that some are better understood in this way while others are not. The Newtonian model of the solar system and linear harmonic oscillators clearly relate to theory and provide cases that support Giere's approach. On the other hand, MS Mouse and the San Francisco Bay model do not fit within any obvious theoretical framework. Philosophers of science who focus on experimentation point out that there are lots of models that are not informed by theory and could not be classified as constituents of a theory (see e.g. Fagan 2013; Leonelli and Ankeny 2013). Further, several of our models stand in some kind of indirect relation to relevant theory. For example, some Neutral Models of Evolution were introduced as a challenge to Evolutionary Theory but are now incorporated into evolutionary thought as a way of understanding the contribution of molecular change to evolution (Wimsatt 2007). Not all models are constituents of, contribute to, or are constrained by theories.

Second, must the models that make up theories be of a particular type, for example mathematical? The cases philosophers of science focus on have an influence over their general accounts of models (and theories). In generating his account of theories as families of models, Giere (see esp. 1988) focused on examples from mechanics. As we see in the next section, this influenced his characterization of models as mathematical and as abstract objects. Theoretical work is carried out in many ways in the sciences. Lorenz's model of motivation and Tversky's model of attention both contribute to theoretical work in psychology and cognitive science, but the relevant theorizing is not mathematically based. Similar demarcation moves are available for characterizing theories as are available when characterizing models. If we define scientific theories as collections of mathematical models, then we have introduced an implicit demarcation criterion. This criterion implies that much theoretical cognitive science and neuroscience is not science. Scientists deploy models relying on many different resources and their theoretical work also relies on many different resources. In the last section of this chapter, we turn to alternate answers to the question, "What are models?" and will find that some of the same issues arise in defending alternate answers to this question as those that arose when attempting to characterize relations between models and theories.

3.4 What Are Models?

We have looked at attempts to classify all the many and varied models into subtypes and looked at attempts to understand models via their relation(s) to scientific theories but can we answer the more direct questions: What are models? or What is a model? Here we look at a few of the answers to these questions and discover that just as there are many different classification schemes for models, there are many different answers to the questions, "What are models?" and "What is a model?"

As we saw in Section 3.2, models can be classified according to their uses or according to their structure. For Weisberg, for example, there are three types of models distinguished by their structure: computational, concrete, and mathematical. Is there anything these three types of models have in common that distinguishes all members of each type as models? Giere (1988), Weisberg (2013, 2007a), and Godfrey-Smith (2006) all agree that we can clearly characterize models and they all share an approach to doing this but disagree on some details of the approach. Let's first look at Giere's (1988) approach to characterizing models.

We saw in the previous section that Giere proposed that theories are clusters of models. Models in this picture are to be understood as abstract objects that are defined by a set of statements and bear a similarity relation with a real system of scientific interest. Giere illustrates this set up in Figure 3.3:

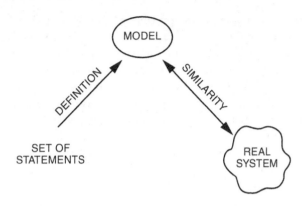

FIGURE 3.3 Giere's model of relationships among sets of statements, models, and real systems. Adapted from Giere (1988), 83.

Giere says that models are abstract entities. The key to understanding a simple harmonic oscillator as a model, on Giere's account, is that "it satisfies the force law $F = -kx$" (1988, 78). The notion of satisfaction here is the same as that used in logic: a set of axioms is satisfied by an object or set of objects. The kind of object that satisfies the force law is not a system in the real world (see also Cartwright 1983). The motion of the bob of a pendulum, for example, does not exactly satisfy the force law. A damped harmonic oscillator better captures a real pendulum, but equations for a damped harmonic oscillator are not satisfied by the bob of a real pendulum. So, for Giere, models are entities that are defined by and satisfy specific equations or sets of equations. Models are distinct from real systems and distinct from the equations that define them. They are abstract entities in the sense that sets are abstract entities. Giere does not say much more about the implied metaphysics here and he shares this stance with many philosophers of science.[9]

Godfrey-Smith hones in on Giere's distinction between models and the sets of statements that characterize them, and he proposes modifications to Giere's diagram and terminology (see Figure 3.4).

For Godfrey-Smith, a model description, which can be an equation, a set of equations or a verbal description, specifies a model system. The model system in turn resembles the target system of scientific interest. These terminological shifts indicate some important

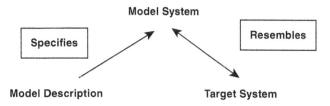

FIGURE 3.4 Godfrey-Smith's proposed modifications to Giere's diagram. Redrawn from Godfrey-Smith (2006), 733 [Modified from Giere (1988)].

differences between Godfrey-Smith and Giere. On Giere's account, models are mathematical models.[10] Godfrey-Smith wants to avoid ruling out nonmathematical modeling as a type of modeling. He has in mind here models in the cognitive sciences, such as Tversky's model of attention, introduced in Chapter 2, which invoke no equations. Godfrey-Smith also wants to draw attention to the special role models play in model-based science. Here he appeals to a distinction Weisberg (2007a) makes between modeling and abstract direct representation (ADR). A model is an intermediary between a model description and a real system used to examine the real system (see also Morgan and Morrison 1999). An ADR is an attempt to directly represent a real-world system of scientific interest. Weisberg says that the periodic table of elements provides a good example of an ADR.

Godfrey-Smith and Weisberg both emphasize the importance of maintaining a clear distinction between model descriptions and model systems. Weisberg (2013, 17) makes critical use of this distinction in arguing against verbal models as a distinct class of models contra Winther (2006). Weisberg argues that verbal models are best understood as model descriptions rather than model systems. The distinction between model descriptions and model systems is easier to maintain in some cases than in others. The DNA model is easy to separate from its model description, but in some models involving equations, it is hard to draw a clean line between equations and model. This issue, among others, is what prompts Jay Odenbaugh's (see e.g. 2018) deflationism about models, which is a rejection of the idea that models are abstract entities of any kind. Odenbaugh argues that models are forms of representation and they "are nothing over and above their mode

48 Characterizing and Classifying Models

of representation" (2018, 11), i.e. the equations, graphs, charts, and so on that scientists use to present their models. Godfrey-Smith is also reluctant to accept Giere's account of models as abstract objects, but Godfrey-Smith is not a deflationist in Odenbaugh's sense; rather, he offers a fictionalist account of models.

Godfrey-Smith argues that model systems are more concrete than abstract mathematical objects. He characterizes model systems as "imagined concrete things" and proposes that we treat them in a similar way that we treat the "imagined objects of literary fiction" (2006, 734–35). Godfrey-Smith argues that his form of fictionalism[11] presents an account of models that better captures scientists' modeling practices than Giere's abstract object account and also applies to more cases of modeling than Giere's account does. Godfrey-Smith's fictionalist account does a better job of accounting for scientists' notions of toy models (Discussed in Teller 2001) and models as caricatures (see e.g. Seger and Stubblefield 1996) than Giere's abstract object account. He also argues that his fictionalist approach provides for a much more amenable notion of resemblance between model systems and target systems, which would bolster Giere's account. (We discuss some alternate resemblance accounts in Chapter 4.)

Models are also characterized as mediating instruments or tools for reconstruction of phenomena (Morgan and Morrison 1999; Morrison 2015). Mary Morgan and Margaret Morrison say their aim as philosophers of science studying modeling is not to provide "well defined criteria for identifying something as a model and differentiating models from theories" (Morgan and Morrison 1999, 12). Rather, they aim to uncover the role of models in scientific practice by looking at a wide range of examples of modeling in different areas of science (see also Morrison 2015). Their notion of models as mediating instruments is derived from this process. We could say that deriving an account of models from the use they are put to avoids the question, "What are models?" This is only true if we have a specific answer to the question in mind. If we take the question to be an ontological one, requiring an answer in terms of specific types of things,[12] then Morgan and Morrison's approach does not answer our question. However, we have seen that saying that models are abstract objects, without going into detail about such objects, perhaps raises more questions than it answers. As we noted at the outset of this section,

Characterizing and Classifying Models **49**

there is a parallel here with alternate approaches to classifying models. We saw that Weisberg aims to classify models by appealing to different structural characteristics, but others classify models by looking at the alternate uses they are put to. Committing to an account of models as abstract objects, or fictions, is not necessarily a superior approach to taking the stance that models are mediating instruments. How we characterize models does inform our approach to understanding their representational capacities and the epistemic work that they do. Each of these topics will be dealt with in separate chapters below, and the way in which answers to the question "What are models?" inform such approaches is important for understanding these issues.

If we adopt one answer to the question, "What are models?", say that they are abstract objects, obvious problems arise for some of our models from Chapter 2. Muscular dystrophy mouse, the DNA model, and the San Francisco Bay model are not abstract objects. Also, none of these models are mathematical. Both the San Francisco Bay model and the DNA model could perhaps be characterized as imagined concrete things but somewhat more concretized than other fictions. The mouse, however, does not seem to fit here. We run into a version of the problem we had when attempting to classify models: If all of our examples of models are to count as models, none of the classification schemes does an adequate job. Here, if all of the examples of models are to count as models, none of the answers to the question, "What are models?" are satisfactory. There are analogous bullet-biting moves available here too (French and Ladyman 1999): models just are those abstract objects that satisfy equations (or a set of equations), and nothing else is a model. This move would remove a large number of our proposed models from contention. Further, as Melinda Fagan points out, to restrict our emphasis to abstract models, as opposed to concrete models, in some scientific fields, such as stem cell biology, puts us at risk of giving "at best, an impoverished, and at worst a perniciously distorted view of the field" (2013, 7).

3.5 Conclusions

Models can be classified in a number of different ways, their relation to theories can be spelled out in different ways and many alternate answers can be given to the question, "What are models?" To best

50 Characterizing and Classifying Models

accommodate the wide variety of scientific practice and scientists' attitudes toward models and theories, perhaps the most inclusive approach possible to these related questions is the best. This approach amounts to endorsing something like "Models are what scientists develop and deploy when they say they are involved in modeling." This kind of practice-driven approach has its adherents in philosophy of science but also has its detractors. We have seen in this chapter that some philosophers provide accounts of what models are that, implicitly at least, rule out much scientific practice as modeling practice. These accounts also rule out much of what scientists say are models as models too. In the chapters that follow we will see that alternate answers to the questions, "How do models represent?" and "How are models appraised?" are often tied to alternate conceptions of models. Our way forward here is to leave all the alternate accounts of models on the table and see how each fares in the context of discussions about representation and appraisal. Alternate accounts are given some support from some approaches to issues about representation but also face different challenges. Similarly, alternate accounts of model appraisal are intertwined with answers to the question, "What are models?"

Notes

1 Daniela Bailer-Jones' (2009) notion of a mechanical model is close to Rosenblueth and Weiner's notion of a material model. She does not use "mechanical model" in the same way that Craver, Darden, or Glennan et al. do.
2 We will see in Chapter 4 that Nancy Cartwright (1999) disagrees here.
3 Biologists are some among many scientists who analyze models and propose classification schemes for them. Philosopher Michael Redhead (1980) has views similar to Lewontin about the nature and role of theoretical models, but his focus is on models in physics.
4 An example of this type of thinking is presented by Jonathan Kaplan (2008), who argues that adaptive landscapes are just mathematical models.
5 All Sober's examples of scientific models are mathematical models.
6 Morgan and Morrison (1999b, 5) are correct to point out that Giere's account differs from semanticists in yet another way: he rejects isomorphism between models and their target systems. These and related issues are discussed at length in Chapter 4.
7 There are more accounts of the structure of scientific theories than the two contrasted here: sets of sentences vs. collections of models. Rasmus Winther (2016) presents and discusses the wide range of such views.
8 Teller's characterization of Giere's approach risks blurring the distinction between the idea that theories are collections of models and the idea that

theories are tools for model construction (see e.g. Suárez and Cartwright 2008). The latter idea is developed as a criticism of the theories as collections of models approach.

9 This relaxed approach to characterizing abstract objects is perhaps what prompts Frigg and Hartman to say "it is not entirely clear what Giere means by 'abstract entities'" (Frigg and Hartmann 2018).

10 Giere (1988, 227–77) does discuss alternate models of sea floor spreading, most of which are not specified by sets of equations, but his discussion of geology is not accompanied by a modified approach to characterizing models. Godfrey-Smith's modifications to Giere's approach easily accommodate geological modeling.

11 There are a range of fictionalist views about models. Nancy Carwright (1983) proposes and defends an influential account, and Arthur Fine (1993) presents some of the history of fictionalist ideas about science. Fine (2009) also criticizes Godfrey-Smith's favored factionalist approach. Roman Frigg (2009) and Adam Toon (2012) each present alternate versions of fictionalism.

12 Hartman and Frigg (2018) catalogue many of the alternate ontological stances on models including physical objects, fictional objects, set-theoretic structures, and equations.

4

MODELS AND REPRESENTATION

4.1 Introduction

There is almost complete consensus among philosophers of science working on models on only one idea: models are representations or models represent. The idea is so prevalent that many do not think it requires supporting argument. Rather, the idea that models represent is the assumed baseline for all discussions of models. In this chapter we examine the idea that models represent and introduce some of the alternate accounts of representation presented and defended. We will also introduce some less common accounts of models and modeling that go against the models-as-representations consensus. In Section 4.2, we introduce the idea that models are representations. Next, in Section 4.3, we assess several influential accounts of representation for scientific models. Finally, in Section 4.4, we examine claims that there are scientific models that do not represent. Issues that arise from understanding models as representations are hard to completely separate from issues that arise from assessing how good (or bad) models are. In this chapter we aim to focus only on representation in Chapter 5, we turn our attention to the epistemological assessment of models.

4.2 Models Are Representations

Paul Teller summarizes the state of play (in the early 2000s) in understanding model-based science as follows:

Models and Representation **53**

> I take the stand that, in principle, anything can be a model, and that what makes a thing a model is the fact that it is regarded or used as a representation of something by the model users. Thus in saying what a model is the weight is shifted to the problem of understanding the nature of representation.
>
> *(2001, 397)*

He adds, "We make something into a model by determining to use it to represent" (2001, 397). So, models are representations and understanding representation is the key to understanding models. Rather than expressing his own specific take on models and representation, Teller was expressing the consensus view in philosophy of science that models are representations. Here is R.I.G. Hughes's expression of the same point: "The characteristic – perhaps the only characteristic – that all theoretical models have in common is that they provide representations of parts of the world, or of the world as we describe it" (Hughes 1997, 325). Finally, Morgan and Morrison say, "models typically represent either some aspect of the world, or some aspect of our theories about the world, or both at once" (1999b, 11). The view that models are representations is almost ubiquitous in philosophy of science and in what follows in this section we will review just a few examples of statements of the view.

Many authors take Giere (see e.g. 1988) to be one of the first to propose a models-as-representations view, but Morgan and Morrison (1999b) point out that such views are found in Mary Hesse's (1966) work, who in turn derived some of her ideas from N.R. Campbell (1920).[1] Hesse expressed the idea that a "physical model is taken to represent, in some way, the behavior and structure of a physical system, that is, the model is structurally similar to what it models" (Morgan and Morrison 1999b, 5). As we saw in Chapter 3, Giere's own view is that models represent an aspect of a real-world system (1988, 2004a). He develops his view via the examination of models presented in science textbooks, such as our pendulum model in Chapter 2. Giere holds that the representation relation between a model and its target is a similarity relation. This constitutes a substantive thesis about representation and we leave the discussion of this and other alternate accounts of the representation relation to the next section. For now, we are just concerned with the idea that models are representations.

54 Models and Representation

Van Fraassen (2008; see also Mitchell 2013) focuses on scientific representations in general. Models are one group of scientific representations and any account of representation is taken to apply to models, along with diagrams, equations, and images. Van Fraassen notes that models, and other scientific representations, are incomplete in that they do not characterize all elements of their targets. For Van Fraassen, this makes scientific representations amenable to the same kind of analysis we use to treat drawings, cartoons, and other kinds of pictures. Cartoons, for example, distort their object while at the same time being successful representations of their object. Van Fraassen says, "It seems then that distortion, infidelity, lack of resemblance in some respect, may in general be crucial to the success of a representation" (2008, 13). He warns against including accuracy in our account of representation. Assessing the accuracy of representations is one approach to understanding representation, but we should not assume that all representation is accurate representation (see Callender and Cohen 2006, 75–76). Morrison puts the point this way: "perhaps the most important feature of a model is that it contains a certain degree of representational inaccuracy" and adds, "it is a model because it fails to accurately represent nature" (2015, 122). Successful scientific representations are often successful in virtue of their idealizations or abstractions (see Wimsatt 2007; Odenbaugh 2018). Both idealization and abstraction involve deliberately inaccurate renderings of a target system. The double helix model of DNA is not an accurate picture of the detailed structure of DNA molecules in vivo but it is a very useful, and highly transportable, representation of a DNA molecule. For Van Fraassen, along with Morrison (2015), the key question facing philosophers of science is not whether models are representations but how it is that abstract, often mathematical, structures can represent phenomena: "What is the pertinent relation that holds or does not hold between the mathematical structure described by our equations and that natural or artificially produced process?" (Van Fraassen 2008, 245). That models are representations is the assumed starting point of this investigation.

We can say a few more things about models as representations before we begin to lay out the details of alternate specific answers to Van Fraassen and Morrison's question in our next section. Giere (1988) contrasts the issues facing philosophers who focus on models with those facing philosophers who focus on theories, construed as sets of sentences. We can ask whether a sentence is true or false and we can construe a set of true sentences as a true theory. However, Giere argues, we cannot attribute truth or falsehood to models, they stand in a different relation to their objects than sentences do.[2] As we saw in Chapter 3, Godfrey-Smith (2006) and Weisberg (2007) develop Giere's approach to modeling by distinguishing between abstract direct representations and models. Abstract direct representations are careful descriptions of an actual system, presented "in order to investigate it directly" (Potochnik 2012, 284). In contrast, modellers aim to "indirectly represent a real-world system by describing a simpler, hypothetical system and investigating that simpler system, in order to draw conclusions about the actual system of interest" (Potochnik 2012, 284). Modellers investigate relations in the model to find out about relations in the world. Models represent the relevant phenomenon but not as an attempt to accurately describe that phenomenon.

Giere (1988) distinguishes between the way models and sentences hook up with the world and Van Fraassen (2008) likens the way models represent to the way drawings and cartoons represent. Each appeals to notions of representation that are applicable beyond the case of scientific models and each introduces insights from approaches to representation more broadly construed. In contrast, Hughes (1997) characterizes the problem of scientific representation, how models represent, as a special problem and one that is separable from general problems of representation. Callender and Cohen (2006) argue that philosophers of science, like Hughes, who think that there is a special problem of representation for science that is distinct from the problem of representation in general are mistaken. They perhaps have in mind something like the following: Hughes introduces a "mathematical problem" – we cannot compare mathematical models to a real system, because the model and the system do not share properties. This problem could be generalized to cover sentences in general: we cannot compare sentences to their objects, because they do not share properties. We do not require that sentences are similar to the states of affairs that they pick out in order for them to be representations and so mathematical models could be representations without sharing any properties with their objects. The idea of sharing properties is only a constraint on a small subset of similarity-based

56 Models and Representation

views of representation (see the discussion of similarity accounts in Cummins 1989) but need not be a general constraint on notions of representation. Not viewing issues of scientific representation within the broader context of discussions of representation in general can mislead. Callender and Cohen (2006) conclude that philosophers of science should use resources from discussions of representation in general. Their own answer to Van Fraassen and Morrison's question is that models represent in exactly the same way that other nonnatural representations – "linguistic tokens, some artworks, pre-arranged signals" – do. Let's now turn to some alternate specific answers to Van Fraassen and Morrison's question given by philosophers of science.

4.3 How Do Models Represent?

4.3.1 Similarity

Van Fraassen, in his earlier work on models (see e.g. Van Fraassen 1980), invoked the notion of models as structures and these structures were related to their target system via a relation of isomorphism. Structuralist, such as structural realists, also defend versions of isomorphism, although most do not require strict isomorphism but require isomorphism between some part of the model and its target system, referred to as partial isomorphisms (see e.g. Costa and French 2003). Giere developed his similarity view of representation partly in response to problems with isomorphism accounts. For Giere, a model represents its target via a notion of similarity.[3] The model is similar, in relevant respects and degrees to its target. Assessing similarity accounts of representation gives a good sense of the challenges facing proponents of specific accounts of representation for scientific models.

Various constraints on answers to Morrison and Van Fraassen's question are proposed. Callender and Cohen ask that answers should clearly respond to the constitution question: "what constitutes the representational relation between the model and the world?" (2006, 68). Giere's response here is that similarity, in relevant respects and degrees, constitutes the model representation relation. Frigg and Nguyen put this point slightly differently, they say that an account of model representation should provide an answer to an epistemic representation question by filling the blanks in an epistemic representation (ER) schema: "M is an epistemic representation of T iff …" (2017, 51). Giere's (1988) similarity account fills the schema as follows: "A scientific model M represents a target T iff M and T are similar in relevant respects and to the relevant degrees" (Frigg and Nguyen 2017, 59). Frigg and Nguyen's alternate candidate answers to the epistemic representation question are presented in terms of necessary and sufficient conditions. This is a clean way for comparing most alternate answers to the question and we will adopt this approach for now. We will revisit this choice later in this section. We can clearly contrast a structuralist answer to the constitution question by filling in an ER schema as follows: assuming a target system exhibits structure S_T and our model is the structure S_M, "A scientific model M represents its target T iff S_M is isomorphic to S_T" (2017, 68). Within this framework, we alter our accounts by adding components to the right-hand side of the relevant schema.

Mauricio Suarez (2003) argues that neither isomorphism nor similarity can ground the representation relation for models (see also Downes 2009). He points to the dismal record for similarity (or resemblance) views of representation in general, citing Goodman's (1976) objections to such accounts, including the counter that while similarity is both a reflexive and symmetrical relation, representation is neither.[4] Frigg and Nguyen capture this kind of consideration in their "requirement of directionality": "models are about their targets, but targets are not about models" (2017, 55). So, an analysis of how models represent must account for this asymmetry. Giere can respond here by pointing out that his similarity account is not a mere resemblance account like this: "A scientific model M represents a target T iff M and T are similar" (Frigg and Nguyen 2017, 58). Rather, Giere's account invokes similarity in relevant respects and degrees. However, this move also is beset by problems. One of which is accidental representation (Suarez 2003; Frigg and Nguyen 2017, 59) via mistargeting. Suarez asks us to imagine that someone dresses up to look like Pope Innocent X. Suarez says that it is a clear case of misrepresentation to go on to suppose that Velazquez's painting of Pope Innocent X represents our dressed up person. Representation does not fail here because similarity fails, as the dressed up person is similar, in relevant respects and to relevant degrees to Pope Innocent X as painted by Velasquez. Giere's similarity view fails to account for misrepresentation cases such as this.

58 Models and Representation

Giere (1999, 2004, 2009) supplements his similarity account with a clause invoking the role of scientists' intentions in presenting a model and/or the use that they put the model to. Rather than shoring up similarity, this move takes the burden off similarity as an answer to the constitution question and appears to characterize the representation relation entirely in terms of scientist's intentions (see Frigg and Nguyen 2017, 61). We deal with this approach, along with similar agents intentions/use-based accounts of representation in Section 4.3.2. Here we turn to Weisberg's (2013) filling out of the similarity account via feature matching.

Weisberg (2012, 2013 Chapter 8) agrees that general accounts of similarity face problems, such as those reviewed above but argues that we can still pursue an account of similarity applicable only in the specific case of relations between models and their targets. Weisberg develops Amos Tversky's (see e.g. 1977) account of similarity, invoked in a psychological account of our judgments of similarity. Tversky's central idea is that similarity judgments are based on feature mapping, what underlies our similarity judgments is the allocation of shared and unshared features. His framework is derived from set theory and Weisberg follows Tversky in using set theory to lay out his formal account of similarity for models and their targets. A model is similar to its target to the extent that it shares features with its target and is dissimilar to the extent that it fails to share features. This idea is formalized as follows: For a set of features, Δ, and a model m, and target t, M is the set of features in Δ possessed by m, and T is the set of features in Δ possessed by t. A weighting function, f, with weights θ, α and β is also assigned (weightings could simply be numbers assigned to the size of subsets of features but other weightings can be considered). The similarity of m to t is given by:

$$S(m, t) = \theta f(M \cap T) - \alpha f(M - T) - \beta f(T - M) \quad \text{(Weisberg 2013, 144)}$$

Weisberg says, "To a first approximation, I think that this is the correct account of the model world relationship" (Weisberg 2013, 144). One way of understanding Weisberg's account is that it provides a formalization of Giere's notion of relevant respects and degrees, a formal spelling out of the right-hand side of Giere's ER schema. One advantage of Weisberg's account over Giere's is that the representation relation, defined this way, is not symmetrical: M can be similar to T to a different degree than T is similar to M (see Frigg and Nguyen 2017, 63). Also, Weisberg's account has a built-in notion of representational accuracy. The upper limit of S(m, t) is the model being an exact replica of its target (see Frigg and Nguyen 2017, 63–64). We look at this aspect of Weiberg's view in our next chapter, as accuracy is one of the candidate ways for assessing models, it is a candidate epistemic virtue of models. So, Weisberg has a similarity account of representation that has some useful features and withstands some key criticisms of similarity accounts. However, when Weisberg expands his account he does so via scientists' goals and interests in picking features of interest and assigning weights. This move does not build on the formal analysis of the model/world relation in kind; rather, it supplements that account with a new type of account, one in terms of scientists' goals and interests. Weisberg can perhaps be understood as presenting a two-part analysis to representation.[5] Where Giere appears to supplant his similarity account with an account in terms of scientific agency, Weisberg supplements his formal account with an account in terms of scientific agency. We now turn to alternate agent-based accounts of representation.

4.3.2 Agent-Based and Use Accounts of Representation

Many philosophers of science maintain that while similarity (or resemblance) has something to do with the model world relationship, it should not play a key role in our understanding of the relationship. Some, for example Teller (2001), argue that this is because a general account of similarity is impossible, and others, for example Van Fraassen (2008), argue that there are too many problems with resemblance relations to make resemblance a key part of our notion of representation. For some this is a specific point about similarity, but others make this point in terms of the possibility of any analysis (or definition or theory) of representation. Both Van Fraassen and Morrison make this latter move. Van Fraassen says, "representation is not to be subjected to definition: it is inexhaustible as a subject" (2008, 31). Morrison echoes this sentiment, saying that we do not need a "philosophical theory of representation" (2015, 7) and that a "theory of representation seems unnecessary" (2015, 155).

60 Models and Representation

Here we need to briefly revisit our acceptance of Frigg and Nguyen's characterization of alternate accounts of representation as alternate ER schema, each presenting different necessary and sufficient conditions for the representation relation. On the face of it, neither Morrison nor Van Fraassen thinks that such definitions can be provided for representation. On the other hand, both Morrison and Van Fraassen, along with Giere (see e.g. 2004b), defend accounts of representation grounded in scientists' agency. For example, Van Fraassen says whether or not A represents B "depends largely, and sometimes only, *on the way in which A is being used*" and continues "*There is no representation except in the sense that some things are used, made, or taken, to represent some things as thus and so*" (2008, 23, italics in original). Perhaps what is going on here is that Morrison and Van Fraassen reject answers to the constitutive question couched in terms of relations between structural features of a model and its target. In other words, they reject what Suarez (2003) calls "naturalistic" accounts of representation, accounts of representation entirely given in terms of mind independent features of a representation and its target. So, rather than rejecting the possibility of any account of representation for models, they reject naturalistic definitions of representation for models. [6] We will proceed as if this is the case. This helps us view agent-based accounts of representation as alternate accounts of representation to similarity accounts and also allows for the characterization of Weisberg's account as a similarity and agent-based hybrid.[7]

Giere (2004b), Morrison (2015), Teller (2001), and Van Fraassen (2008) all present and defend accounts of model representation in terms of scientists' use of their models. We introduced Van Fraassen's expression of his version of the approach above. Giere sums up his approach as follows: "It is not the model that is doing the representing; it is the scientist using the model who is doing the representing" (2004b, 747). Morrison concurs with Van Fraassen (and presumably Giere) on this point, saying that she agrees with Van Fraassen that representation occurs only when mathematical artifacts, such as models, are "used in a certain way or taken to represent things in a specific way. In that sense there is no notion of a representation that exists in nature without some accompanying cognitive activity" (Morrison 2015, 125). Morrison supports the agent-based view of representation in part because it helps us make sense of the following situation: many different phenomena can be represented using the same mathematical model. Our logistic growth model is a nice example here, the model can be applied to the growth of bacterial populations just as readily as it can be applied to the growth of human populations. Morrison says that this situation renders it "obvious that the representation relation must involve the kind of indexical element that Van Fraassen suggests" (2015, 125). Further, this situation leads Morrison (along with Van Fraassen) to adopt an approach to understanding modeling (and representation) that is highly context-specific. Her approach to philosophy of science is to focus on examples of scientists' modeling practice in detail. We find out about representation by finding out about the actions of the scientists deploying specific models.

Morrison and Van Fraassen put scientists as agents front and center in their account of representation, but at the same time they both also express great interest in why it is, and how it is, that mathematics is the top modeling (and representational) resource in science (see e.g. Morrison 2015, Chapter 2). This indicates that they each believe that perhaps there is something inherent in mathematical models that contributes to their usefulness as scientific representations. Examining the way in which given scientists use a mathematical model will likely not fully address this issue. After all, no amount of careful work on behalf of an individual scientist will render a shovel or a hammer a better model of population growth than a logistic growth equation-based model.[8] Scientists make choices about the most apt modeling resource for the type of dynamical system they confront in the world and as we noted in Chapter 2, entire textbooks are dedicated to the construction and selection of appropriate models for the phenomena of interest. Morrison is right that a focus on the choices that scientists make in specific contexts is crucial to our understanding of modeling but it appears that the use scientists put models to cannot entirely account for their representational capacity. The agent-based, use, or some call it the pragmatic approach to representation should clearly inform our understanding of the model world relation but likely needs supplementation to give a full account of this relation. Some concede that scientists' agency is part of the picture but present hybrid accounts of representation, which include other aspects of the model-world relation.

62 Models and Representation

As we noted above Weisberg's account can be construed as a hybrid account. Weisberg proposes that the model-world relation is best analyzed via his feature matching account of similarity, but he also aims to take into account scientists' aims and goals in deploying models. Weisberg points out that "scientists are often interested in comparing the relationship that the model actually holds to the world to the one that they are interested in achieving between the model and the world" (2013, 150). He also says that his account "reflects judgments about the relationship of models to their targets that scientists actually make" (Weisberg 2013, 155). As noted above, Wendy Parker (2015) sees a problem here. Where we suggest Weisberg has a hybrid account, Parker sees a conflation of similarity relations between models and their targets and scientists' judgments about how their models match up to their targets. Weisberg has not given up on similarity, as Giere has, but he is perhaps not clear enough on the exact role that he takes scientists to play in shoring up the relation between models and the world.

Callender and Cohen (2006) provide an account of representation for models that is close to the agent-based account but includes an additional component. As we noted above, they say that models represent in the same way that "linguistic tokens, some artworks, prearranged signals" and so on do. They say, "the varied representational vehicles used in scientific settings (models, equations, toothpick constructions, drawings, etc.) represent their targets (the behavior of ideal gases, quantum state evolutions, bridges) by virtue of the mental states of their makers/users" (Callender and Cohen 2006, 75). This part of their account appears to be a pure agent-based or use account. Frigg and Nguyen use the term "stipulative fiat view" to label Callender and Cohen's account: "A scientific model M represents a target system T iff a model user stipulates that M represents T" (2017, 55). This part of Callender and Cohen's account of representation for scientific models is consistent with Giere, Morrison, and Van Fraassen's use-based account, and Morrison recognizes this (2015, 130). However, Callender and Cohen explain the representational power of scientific models, along with other nonnatural representations, in terms of fundamental representations: mental representations (see also Odenbaugh 2018). The full account of representation for models is an account of derived representation and the full account also requires spelling out the fundamental representation relation of mental states from which models

Models and Representation **63**

derive their representational capacity. Callender and Cohen take the stipulative fiat account to do most of the work in the case of scientific representations but only if this account is shored up by an account of fundamental representation. This renders their account explicitly at odds with both Morrison's and Van Fraassen's accounts, as Morrison and Van Fraassen both reject the possibility of mental representation.

The above accounts of the model target relation are just a few of the huge number of alternatives (see Frigg and Nguyen (2017) for a larger sample of alternate approaches). Work on how models represent continues to proliferate and, to date, there is no consensus on how models represent. We said at the outset of this chapter that there was consensus on the claim that models are representations, but there are a few philosophers of science who reject this consensus view. We turn to some representatives of this group in the next section.

4.4 Modeling without Representation

Above we have examined the consensus view that models are representations. The strongest version of this view is that all models are representations and all successful modeling in science is achieved via representation. No direct negation of this strong claim is defended in philosophy of science although one approach looks to come close to doing so: William Wimsatt's (2007) treatment of "false" models. He says models are "mere heuristic tools to be used in making predictions or as an aid in the search for explanations" (2007, 94) and goes on to say that we use models because they are effective tools. Looking more carefully at Wimsatt's treatment of "false" models reveals that he claims that all models misrepresent and not that models do not represent. As we saw above, misrepresentation and nonrepresentation must be kept distinct. Many philosophers of science, such as Morrison, agree with Wimsatt that models misrepresent. Wimsatt's instrumentalism about models is introduced to handle the problem of misrepresentation and not as a nonrepresentational account of all models. Several philosophers of science defend the weaker claim that some models do not represent or that some nonrepresentational aspects of otherwise representational models do epistemic work. In this section we present and unpack some examples of these claims about nonrepresentational models.

64 Models and Representation

We first look at two alternate distinctions between representational and nonrepresentational models: Nancy Cartwright's (1999) and Evelyn Fox Keller's (2000) (developed by Emanuele Ratti (2018)). Later in the section we consider two additional accounts of nonrepresentational models due to Isabelle Peschard (2011) and Till Grüne-Yanoff (2013).

Cartwright (1999) argues that theories do not represent and nor do models constructed from theories, which she calls interpretive models. She reserves the term representational models (alternately called phenomenological models) for models of actual systems. We can use one of our models, the harmonic oscillator model, to illustrate Cartwright's distinction. Giere (1988) takes the harmonic oscillator to model the horizontal component of motion of a real-world pendulum. Cartwright disagrees, she says that such a model does not fit the real pendulum even approximately and further, that the theory such models are derived from contains no guidelines for constructing a representational model of a real pendulum (cf. Morrison 2015, 133). According to Cartwright, the harmonic oscillator, and analogous models, concretize relations between abstract concepts and as such, are interpretive models. They facilitate our understanding of theoretical relations but are not best understood as representations of real systems, such as pendulums.

Where Cartwright distinguishes between models derived from theories and those designed to represent real systems, Fox Keller (2000) distinguishes between "models-of" phenomena and "models-for" intervention and manipulation. For Cartwright, nonrepresentational models are theoretically involved models, but Fox Keller's distinction is between different models of real systems, some of which are representational and some are "tools for material change" (Ratti 2018, 2). Cartwright and Fox Keller find nonrepresentational models at opposite ends of the theory–experiment continuum.

Ratti (2018) adopts and develops Fox Keller's distinction between models-of and models-for. Our models of thirst (see Chapter 2) are mechanistic models representing aspects of the thirst process. Such models are often presented in an attempt to explain the relevant phenomenon and some, such as Craver (2009), have argued that the more accurate the representation, the better an explanation the model provides. Ratti distinguishes between models-of and models-for in this way:

> The components of the model are used with different purposes in mind; when models are considered 'models of', components are considered as parts of an explanation, unlike in models as 'models for' where components are seen as potentially triggering specific effects in other contexts.
>
> *(2018, 8)*

Models-for introduce new ways of "manipulating biological entities" and can involve the repurposing of a biological mechanism to new ends. CRISPR-Cas systems provide a nice example of models-for. CRISPR-Cas systems were first discovered operating in the immune system of microorganisms. CRISPR-Cas9 was introduced as a model of how these organisms' immune systems work. The system works in part by splicing and editing DNA. The CRISPR-Cas9 model-for is also used for human DNA editing and this is achieved by repurposing a biological mechanism for new ends. One way in which we assess the success of CRISPR-Cas9 as a model-for is via its "portability," "the ease with which components of mechanistic models can be used and combined to solve problems in a different experimental system" (Ratti 2018, 22). We do not assess CRISPR-Cas9, construed as a model-for, in terms of its representational capacities or its potential for explanation of relevant phenomena. We now have two candidate groups of nonrepresentational models: models-for and interpretive models. A look at some other examples of modeling work reveals more examples of nonrepresentational models that do not fit neatly into either of these groups.

Isabelle Peschard (2011) focuses on the constructive use of models: the role that some models can play in the development of new models. Peschard argues that we should focus on the use models are put to by scientists as epistemic tools. She examines modeling work on wakes in fluid dynamics. Although models of wakes are representational, she points out that their most successful role is in the development and construction of new models. For example, once the model of a wake behind a single cylinder moving through fluid is developed, its primary use is in generating new models of multiple wakes. We can

66 Models and Representation

illustrate her point about model generation via one of our examples: the DNA model. No one is likely to argue that the DNA model does not represent a DNA molecule, but the DNA model, as we pointed out in Chapter 2, is also foundational in model construction in molecular biology. Peschard refers to this as the "generative constructive use" of a model and her point is that this aspect of models is nonrepresentational. There is a parallel here between Peschard's claim and Ratti's, in developing Fox Keller's distinction: both point to two aspects of one model that are important in scientific work and argue that the nonrepresentational aspect of the model has a crucial epistemic function.

Grüne-Yanoff (2013) also defends a role for nonrepresentational models by looking closely at the epistemic role various models play. He says that if models are taken to be representations, then good models are good representations. He surveys some of the ways of filling out the implied notion of representational adequacy here, for example, ones derived from the similarity approach reviewed above. According to Grüne-Yanoff, the similarity view is committed to the idea that we learn from models to the extent that the model captures aspects of the real world or system we are interested in (2013, 850). Grüne-Yanoff counters that looking at scientific practice reveals that scientists "can learn from a model even without establishing its representational adequacy" (2013, 852). He rejects Wimsatt's "mere heuristic tools" alternate to representationalism arguing that this approach does not help us to grasp how nonrepresentational models are appraised. His alternative approach to model appraisal is that scientists learn from a model to the extent that their confidence in hypotheses of interest change as a result of examining or using a model. On this account, the Schelling model of segregation lowers our confidence in the racism hypothesis for segregation (Grüne-Yanoff 2013, 857). After considering a number of models, including the Schelling model, Grüne-Yanoff concludes that many established practices for appraising models are set aside by representationalists.

The claims that a subgroup of models, or an aspect of one model, are nonrepresentational are generated by a close look at scientific practice and the ways in which scientists develop, deploy, and appraise models and the context in which they do this work. Those committed to the view that all models represent, and that all successful modeling is achieved via representation, can challenge the nonrepresentationalists by looking in detail at the relevant scientific practice. Morrison, for example, does this to formulate a response to Cartwright. Morrison (2015) argues that a close look at the relevant physics reveals that Cartwright's interpretive models are representational after all. This debate is still ongoing. The various nonrepresenationalist approaches could be countered in a piecemeal fashion or representationalists could concede that some models, or some aspects of models, are nonrepresentational. Conceding this point opens up the possibility for alternate approaches to model appraisal that are not tied to a notion of representation. We turn to alternate approaches to model appraisal in our next chapter.

4.5 Conclusions

There are many alternate ways of cashing out the model–target relation and we have examined a selection of them here. While there is near consensus on the idea that models represent, we have seen that there is no consensus on how the model–target relation should be characterized. Many more ways of filling in Frigg and Nguyen's ER schema have been proposed than those reviewed here (see their 2017 for details) and debate is still ongoing over the virtues of these alternative accounts. Many of these alternate accounts of representation for models are developed via detailed studies of scientific practice. Philosophers of science have found cases of modeling that look to support one or other account of representation, but we also found that some argue that an invoking of nonrepresentational models better accounts for some aspects of scientific practice.

Our discussion of so-called "nonrepresentational" models gives an indication of the difficulty in disentangling the issue of how models represent from the issue of the role representation plays in making models better (or worse). In this chapter, we have tried to keep the focus squarely on the representation issue, but a further brief discussion of one of our accounts of representation illustrates how difficult this can be. Several people, including Weisberg himself, have argued that Weisberg's feature matching account of similarity is best suited to treat the issue of model accuracy (Weisberg 2015; Parker 2015; Frigg and Nguyen 2017). Recall, that the upper limit of Weisberg's match between model and target is the situation in which they are the same

68 Models and Representation

(Weisberg 2013, 147). Emphasizing this feature of the account implies that the account is best designed to formally fill in the degree of similarity between a model and its target, in other words, presenting an account of the accuracy of the model. An answer to our constitutive question need not invoke accuracy and, as we have seen, some argue that a satisfactory account of the representation relation must cover cases in which models are prima facie inaccurate representations (Callender and Cohen 2006, 75–76). Accuracy of representation could be one of the virtues of a scientific model, but such virtues are epistemological and a discussion of these aspects of models is a discussion of the epistemic appraisal of models. In our next chapter, we consider several of the alternate approaches to the epistemic appraisal of models.

Notes

1 Van Fraassen (2008) traces the view back to Boltzman (1905, 1911).
2 Laura Perini (2005) disagrees that sentences are the only appropriate vehicles of truth and falsehood.
3 Susan Sterrett (2017) provides a history of the notion of physically similar systems that helps motivate the choice of the account representation as similarity, particularly among philosophers of science who focus on physics.
4 Robert Cummins (1989) summarizes the problems for similarity-based accounts of mental representation.
5 Wendy Parker (2015) has a different take here: she argues that Weisberg does not clearly distinguish between formally analyzing the representation relation in terms of similarity and understanding scientists' judgments of similarity.
6 Understanding the situation this way is helped by the fact that both Morrison and Van Fraassen explicitly reject naturalistic representation relations and also both appear to reject the idea of mental representation, central to cognitive science (Van Fraassen 2008, 24; Morrison 2015, 125).
7 Understanding the situation like this also allows for consistency with the way alternate accounts of meaning are treated in philosophy of language. Use accounts of meaning are just as much accounts of meaning as those given in terms of reference or some other relation between sentences and their objects.
8 Callender and Cohen counter here, saying that choosing between the shovel and the logistic equation is an entirely pragmatic matter and has nothing to do with their "representational properties *per se*" (2006, 75).

5

WHAT MAKES FOR A GOOD (OR BAD) MODEL?

5.1 Introduction

Ecologist Amy Hurford (2012) proposes many reasons for scientists to produce models, including the following: Scientists make models to make a quantitative prediction or to make a qualitative prediction. They use information from one scale to understand another (i.e., multiscale modeling) or they use models to clarify the reasoning of a given argument or to investigate hypothetical scenarios. Models can be used to motivate experiments or to disentangle multiple causation. Models can be used to make an idea or hypothesis precise or to characterize all theoretically possible outcomes. Wimsatt (2007) also provides a list of very similar purposes for introducing models in science. Given that models are proposed for many purposes in the sciences, it is likely that there are varying methods for evaluating models and varying grounds for appraising them. One group of scientists can prioritize models that best match relevant empirical data while others prioritize models with very little apparent fit to such data. Models are valued for their predictive power, their explanatory power, their productivity in hypothesis generation, their troubleshooting capabilities and, in some cases, their similarity to a real-world system. There is often no one way in which a given model is considered good (or bad).

70 What Makes for a Good (or Bad) Model?

In this chapter we present and assess some approaches to model evaluation and appraisal proposed both by scientists and philosophers of science. Given the huge range of purposes for models and the implied huge range of alternate approaches to model appraisal, in this chapter we select just a few of the issues in model appraisal that have attracted philosophical attention. In what follows we will consider assessing the extent to which models provide accurate descriptions of real systems; the extent to which models are confirmed or fit relevant data; and how well models explain phenomena. We will also look at trade-offs between model aims and hence methods of appraisal and, finally, we will consider model robustness.

5.2 Accurate Description

In Chapter 4, we introduced the prevalent idea that models represent. Representations can be appraised for their accuracy. While some require that standards of accuracy be built into an account of representation (see e.g. Frigg and Nguyen 2017), here we consider issues of the accuracy of models separately.[1] One way in which models are appraised is by asking whether they present accurate descriptions of their target system. As Lloyd says, a variety of evidence can increase our confidence in "the accuracy of any particular description of a natural system by a model" (2017, 924). Some of our models appear to clearly ask for this kind of appraisal: the San Francisco Bay model; the DNA model; and the detailed mechanistic model of thirst in mammals. Ratti (2018) sets up the models-of/models-for distinction by characterizing models-of as accurate representations of phenomena. Craver (see e.g. 2009), along with many philosophers who focus on mechanisms, argues that mechanistic models must describe a real causal structure invoking real components. For Craver, mechanistic models undergird explanation. Ratti puts the point like this:

> The gold standard for a mechanistic explanation is that it should include all the relevant features of the mechanism we investigate. This is what distinguishes a 'complete' and adequate mechanistic explanation from a mere how-possibly model or a mechanistic sketch.

(2018, 6)

What Makes for a Good (or Bad) Model? **71**

Our detailed model of thirst is arguably a mechanistic model that would satisfy Craver's standards of completeness and adequacy. We leave issues of explanation for a separate section and here focus just on the idea of appraising models' representational accuracy.

We saw in Chapter 4 that Weisberg's feature mapping similarity account of representation has a built-in maximal notion of similarity, which is an exact match between elements of the model and elements of the target phenomenon. Weisberg introduces a distinction between calibrated and uncalibrated modeling, which fits well with his feature matching approach to similarity: "In calibrated modeling, the goal is to make a high-fidelity model of a target" (2013, 94). In contrast, in uncalibrated modeling the aim is to produce a "qualitative fit between model and target. The idea is to see if we can more or less reproduce the phenomenon of interest using a model." Weisberg says that the San Francisco Bay model is a calibrated model. The model is a highly accurate scaled model of the San Francisco Bay and shares many similar properties with the San Francisco Bay. The DNA model, on the other hand, could be thought of as either a calibrated or an uncalibrated model. With respect to bond angles between molecules and other key features of molecular structure, the DNA model can be seen as an attempt at a high-fidelity model of the relevant phenomenon. On the other hand, the use the model is put to in generating other models in molecular biology (cf. Peschard 2011 on models of wakes), highlights the uncalibrated aspect of the DNA model. Understood from this perspective the DNA model "more or less reproduces the target of interest."

For Craver, successful models are accurate models of the relevant target and for Weisberg, successful, calibrated, models are high-fidelity models of their targets. Clearly accuracy is important in some aspects of modeling, but there are limits to using accuracy as the main mode of appraisal for models. First, as Peschard (2011) and Ratti (2018) point out, the same model can have different aspects and be put to different uses. Let's stick with the DNA model example. We can assess both the accuracy of the DNA model and its usefulness in generating models in molecular biology. If we push our accuracy assessment, the DNA model starts to founder. Certainly, if we place the kind of standards of accuracy Craver requires of a mechanistic model on our DNA model, it does not pass muster. For example,

72 What Makes for a Good (or Bad) Model?

the DNA model has no resources for accounting for the folding and unfolding of DNA, DNA methylation, and the other host of cellular processes that go into producing the familiar patterns of transmission and transcription characterized by molecular biologists. The DNA model is one of the keys to unraveling all of these cellular processes, but the model is nowhere near a high-fidelity model of such processes. Second, when accuracy is understood via a notion of similarity, the limit case for accuracy is sharing all the same features in all the same relations to one another. However, a life size sat-map of my neighborhood may be accurate, but it strains our idea of a map, just as a detailed high-fidelity model strains our idea of a model (see Van Fraassen 2008, 20). Also, if our thirst model was too finely calibrated to mice, it would start losing its applicability as a model of mammalian thirst, capturing fine-grained accuracy trades-off with the portability (Ratti 2018) or generality (Matthewson and Weisberg 2009; Michael Weisberg 2013) of a model. (We will look at this and other trade-offs in a bit more detail in Section 5.6.) Third, some models appear to provide no starting point from which to make assessments of accuracy. The Edgeworth Box, for example, appears to lie at the far end of the calibrated/uncalibrated model continuum. The model allows us to specify points of interaction between each agents' choices over use of the available resources, but it does not come anywhere close to an accurate description of interactions between, even economic, agents. However, the model is very useful in economics and is foundational in the producing of other economic models. Such highly idealized models, perhaps highly uncalibrated models, call out for different methods of appraisal. Finally, one person's highly idealized model is another's "work of fiction" (Cartwright 1983, 153), wrong model (Box and Draper 1986, 424), or false model (Wimsatt 2007). Wimsatt counts idealization as one of the ways in which models can be false, along with their incompleteness and misdescription of relevant variables (2007, 101–2). If some models are all wrong, are false, or are fictions, then clearly we need alternate modes of epistemic appraisal than descriptive accuracy.

We will consider how appraisal of accuracy interacts with other forms of appraisal, such as whether or not a model provides a good explanation, in the following sections. Next we turn to the issue of model confirmation and the related issue of a model's fit with data.

5.3 Confirmation and Fit

One traditional way of appraising models is to assess the extent to which they are confirmed. One way a model can be confirmed is by fitting its target or fitting relevant data. Understanding model fit to data is easier than understanding model fit to targets or real systems and so we will start with the first case. We can illustrate the idea of models fitting data using simple statistical models and also the Lotka-Volterra model. If the graph of our data, such as plant height measurements, closely fits the graph produced by our model, our model fits the data. There are many tools in statistics for assessing curve fit and such tools are used in appraising the relevant models, which Lloyd refers to as "ordinary statistical techniques of evaluating curve fitting" (Lloyd 2017, 922; see also Odenbaugh 2018). The Lotka-Volterra model predicts a pattern that can be graphed (see Figure 2.8) and fluctuating population patterns of hare and lynxes can also be graphed (see Figure 2.9). The final 20 years or so of hare and lynx population fluctuations are somewhat close to the model, as assessed by matching the relevant curves, but the previous 60 years do not closely fit the model. On fit grounds alone, the hare/lynx data does not strongly confirm the model (Michael Weisberg 2013, 93). Lloyd connects this notion of fit to model–target fit. She says that the model matches the natural world system as a result of the match between a curve derived from a model and "(data models extracted from the) natural system" (2017, 923). So, one source of confirmation for models is model fit.

Lloyd proposes several other ways models can be confirmed in addition to model fit in both evolutionary biology and climate science. These include independent testing of model assumptions; range of instances of fit between the model and a variety of systems; and robustness (Lloyd 2009; 2017). (We deal with model robustness in Section 5.6.) One way of looking at Lloyd's approach is as a pluralistic approach to confirmation as a method of model appraisal.[2] Parker (see e.g. 2009) and Winsberg (2018), following Parker, challenge Lloyd's approach to model appraisal, specifically in climate science.[3] Parker's challenge is that climate scientists do not assess how well confirmed their models are; rather, they assess their models' adequacy-for-purpose. Winsberg argues that rather than assessing the extent to which models are confirmed, "Evaluating model skill[4] is the central

74 What Makes for a Good (or Bad) Model?

topic in the epistemology of climate models. This is not a matter of determining the 'correctness' or 'fit' of climate models, but rather their suitability for various prediction and projection tasks" (2018, 173). What is adequacy for purpose? Winsberg says that when we answer the question "What *it is* for a climate model to be a good one?" we need to answer "To be a good model is *purpose relative*" (2018, 33; Parker 2009). He goes on to say that as most climate models are used for climate prediction, they should be assessed to the extent that they are adequate to this purpose (2018, 41).

In the previous section we saw an approach to model appraisal, assessing descriptive accuracy, criticized on the grounds that particular models were not amenable to such an approach. Here, in contrast, the challenge comes from the perspective of the epistemic outlook of the relevant scientists. Parker and Winsberg's challenge is that climate scientists do not assess the extent to which their models are confirmed. Lloyd also works closely with climate scientists, but she does not support her case for confirmation via a study of climate scientist's practice. Rather, her proposal is couched as a criticism of what she takes to be more limited views of what confirmation amounts to proposed by other philosophers of science. This debate reveals some of the different range of philosophical issues that are brought to the fore by studying model-based science. One such issue is a methodological one in philosophy of science. Some philosophers are focused on structural relations between models and their targets and also on applying formal methods to characterize these relations. Others are focused on accounting for a wide variety of aspects of scientific practice, including our current case of local scientific practices of epistemic appraisal (see Thomson-Jones 2010). Winsberg endorses this latter view saying

> philosophers do better to paint a picture of the consensus of the scientific community, based on features of that community's social organization, than to try to provide a normative framework from which we can demonstrate the reliability (or its absence) of such-and-such modeling result.
>
> *(2018, 162; cf. Solomon 2001)*[5]

Winsberg does not only practice social epistemology, as he implies here, he also tackles other epistemological issues, such as more

What Makes for a Good (or Bad) Model? **75**

traditional approaches to prediction, but his emphasis on what philosophers can learn from scientists' social practice is clear. There is room for combined approaches in philosophy of science that highlight formal aspects of modeling and model appraisal as well as approaches focused on scientific practice and scientists' social organization. In the next section we examine the role of explanation in model appraisal. Alternate methodologies in philosophy of science have also led to different analyses of explanation.

5.4 Explanation

Models can be appraised on the extent to which they provide a good explanation or by the part they play in an explanation. This means that the discussion about what makes a good model can be informed by views about goodness of explanation, and as a result, discussion of these issues in philosophy of science is influenced by differing accounts of explanation (see e.g. Bokulich 2011). To begin our discussion, let's return to Craver's approach to models introduced in Section 5.2.

Craver argues that for a model to provide an explanation of a phenomenon, the model must describe the mechanisms underlying the phenomena (see e.g. 2009). Alisa Bokulich (2011) characterizes Craver as invoking the constraint on explanation that explanatory models must characterize the relevant phenomenon correctly and completely in terms of its underlying mechanisms. Here Craver sides with one school of thought on explanation in philosophy of science: an explanation must trace a detailed causal chain that accounts for the phenomenon requiring explaining (see Bokulich 2017). He also leans heavily on the idea that a model must provide an accurate description of relevant phenomena. We have seen that lots of scientists do not aim at accurate descriptions when developing their models. Perhaps this means, following Craver, that models not providing accurate descriptions cannot provide explanations or form part of explanations. Alternately, different models could provide different kinds of explanations. For example, our models from population genetics or from evolutionary game theory specify no underlying causal mechanisms, but evolutionary biologists argue that such models are explanatory and that we can appraise alternate models according to their explanatory adequacy (see e.g. Matthewson and Weisberg 2009, 188). On this

76 What Makes for a Good (or Bad) Model?

account, such models are explanatory to the extent that they reveal underlying patterns shared by related phenomena. Here philosophers of science are divided. Some take the line that revealing underlying patterns without uncovering a detailed causal story is not an explanation. This is how Bokulich (2017) understands Craver's position. Some say that explanations in population genetics (and ecology) are not full explanations, they are "how possibly" explanations as opposed to "how actually" explanations (see Forber (2010) for more on this distinction). Finally, some defend accounts of explanation that include population genetics models and evolutionary game theory models as explanatory. We turn now to this latter position via a discussion of the role of idealization in explanation.

Bokulich, in contrast to Craver, says that models are "by definition incomplete and idealized descriptions of the systems they describe" (2017, 104). The questions she asks are whether or not such models can explain and what role idealizations play in these explanations. Before turning to these questions, a brief word on idealization is in order. Weisberg says that idealization is "the intentional introduction of distortion into scientific representations" (2013, 98). He goes on to distinguish three types of idealization: Galilean idealization, minimalist idealization, and multiple-models idealization. Galilean idealization introduces distortions into models to simplify them and make them more mathematically tractable. Weisberg notes, following Ernan McMullin (1985), that "Galilean idealization takes place with the expectation of future de-idealization and more accurate representation" (2013, 100). In minimalist idealization, the modeller introduces only core factors that give rise to the phenomenon of interest. On this approach modellers produce minimal models, which are often very simple models that do not aim to capture much in the way of detail about the phenomenon of interest (see Batterman 2002; Batterman and Rice 2014). Modellers develop minimal models with no expectation of future de-idealization. Finally, multiple-models idealization is the "practice of building multiple related but incompatible models, each of which makes distinct claims about the nature and causal structure giving rise to a phenomenon" (Michael Weisberg 2013, 103). We can now return to Bokulich's questions about the relation between idealized models and explanation.

Craver's view directly entails that models containing idealizations cannot explain. In a related approach to explanation, Michael Strevens (2008) also denies an explanatory role to idealized components of models. Strevens holds what Bokulich calls "the received" view of model explanation, which she characterizes as follows: "it is only the true parts of the model that do any explanatory work. The false parts are harmless, and hence should be able to be de-idealized away without affecting the explanation" (2017, 108). Strevens understands idealizations as "harmless" falsehoods whose "role is to point to parts of the actual world that do not make a difference to the explanatory target" (2008, 318). This approach is countered by Bokulich (2011), Batterman and Rice (2014), and Wimsatt (2007), among others, all of whom hold a version of the view that it is because of their idealizations (or falsehoods) and not in spite of them that models explain. Bokulich says that idealizations in some models are ineliminable and "do real explanatory work" (2017, 108). One way of understanding this dispute is to take Strevens to be referring to Galilean idealizations while Bokulich, Batterman, Rice, and Wimsatt are referring to minimal idealizations, but this does not make the issue go away. Strevens can respond to Batterman, for example, that a minimal model is an explanatory model, only to the extent that it has a causal explanatory core as along with its idealizations (cf. Michael Weisberg 2013, 101). Another way past this debate is to reframe it in terms of scientists' aims and purposes in introducing their models. Here, following Parker and Winsberg's approach introduced in the previous section, we look to the relevant scientist's practice of assessment of explanations for guidance. Taking this approach requires philosophers of science to take a population geneticist seriously when they say that their model is explanatory and make an effort to understand what sense of explanation the scientists are invoking. Here, again, we see a contrast in methodology in philosophy of science. Critics of the approach that takes scientific practice seriously can say, "We aren't really interested in what scientists say, especially if they invoke an incorrect notion of explanation." The correct account of explanation is the critics' favored account in the philosophy of science. One risk with this response is the possibility of another kind of demarcation problem, analogous to those we discussed in Chapter 3: We adopt an account of explanation for models, which implies that a large number

78 What Makes for a Good (or Bad) Model?

of scientific models are not explanatory (say all models in population genetics). This in turn implies that the relevant science does not provide any explanations. There is room for biting the bullet here but more philosophers of science appear to prefer pluralism about explanation than biting this bullet. We now turn to trade-offs between alternate desirable epistemic properties of models.

5.5 Trade-Offs

Scientists, and philosophers of science, value models that are general. For example, the Newtonian model of planetary motion, is valued for its wide application to many systems of interacting massive objects not just our solar system. The Lotka-Volterra model applies to many predator–prey systems in biology and the Edgeworth Box applies to all pairs of agents confronting choices over two fixed resources and so both have generality. However, many models are valued for the precise way in which they capture a particular phenomenon, for example, Craver values mechanistic models that provide detailed, accurate descriptions of their target. Intuitively, general models cannot be precise models and vice-versa and increasing generality implies reduced precision and vice versa. Population biologist Richard Levins (1966) argued that trade-offs between model desiderata, in ecology and population biology, could be formalized and certain desiderata for models could not both be "simultaneously maximized" (Matthewson and Weisberg 2009). Levin's idea faced a strong challenge from Orzack and Sober (1993), who argued that Levin's desiderata did not trade-off in the way he suggested. Subsequent work in philosophy of science has gone some way to vindicating Levins but not without criticism and refinement of his original proposal (see e.g. Odenbaugh 2003; Matthewson and Weisberg 2009). Here we look at John Matthewson and Weisberg's arguments that model precision and model generality do indeed trade-off. We also recast Ratti's point about relative portability of models-for as a trade-off between model desiderata.

Weisberg and Matthewson (2009; see also Weisberg 2013) set out to establish formally that model generality and precision trade-off. For them, model generality "is a measure of how many phenomena a model or set of models successfully relate to" (2009, 180). Precision is understood via the idea that "changes in precision will effect the set

What Makes for a Good (or Bad) Model? **79**

of models picked out by a description. More precise descriptions will pick out subsets of the sets of models picked out by less precise descriptions" (2009, 179–80). They distinguish two types of generality: actual generality (a-generality) and possible generality (p-generality). These account for the difference between actual and possible model targets. Matthewson and Weisberg do not find simple trade-offs between precision and generality across the board; rather, they argue that model set p-generality can only be increased if precision is decreased and model set a-generality and precision cannot be simultaneously increased. They conclude that these specific attributes of sets of models trade-off. This conclusion supports our intuitive idea that generality and precision trade-off, but careful work on formalizing the relevant trade-offs, backed up with a study of examples in population biology, reveals that trade-offs of model desiderata are not as widespread as our intuitive picture indicates, nor are they as widespread as Levins thought.

One interesting further conclusion of Matthewson and Weisberg's analysis relates to our discussion of explanation in the previous section: an increase in model generality is *"ceteris paribus*, associated with an increase in explanatory power" and increased precision is associated with decreased explanatory power (2009, 189). This is not a direct result of their formal analysis; rather, they draw the conclusion based on biologists claims about the relative explanatory merits of heir models. This conclusion is in tension with both Craver's and Strevens' accounts of explanation.

So, some trade-offs between model desiderata can be established formally for particular sets of models, but there are other less formal trade-offs between model desiderata revealed by close scrutiny of scientific practice. In Section 4.4, we introduced Ratti's use of Fox Keller's models-of/models-for distinction. As we noted, Ratti says that the success of CRISPR-Cas9 as a model-for is via its "portability," "the ease with which components of mechanistic models can be used and combined to solve problems in a different experimental system" (2018, 22). In contrast, understood as a model-of, CRISPR-Cas is a high-fidelity representation of part of bacterial immune response. An accurate mechanistic model such as this is hard to decontextualize and as a result lacks portability. Ratti argues that whether a model is considered a model-of or a model-for depends on scientists' cognitive

80 What Makes for a Good (or Bad) Model?

dispositions towards the model. We could characterize the trade-off between portability and accurate representation as a trade-off between model attributes of different aspects of a model resulting from the different purposes scientists attribute to the model. Trade-offs between the epistemic virtues of our models can be formally demonstrated or be discovered by close scrutiny of scientific practice and scientists' attitudes.

5.6 Robustness

Levins (1966) introduced the notion of robustness for models in the paper introduced in Section 5.5. For Levins, when we have multiple, distinct and yet similar models of the same phenomenon "[I]f these models, despite their different assumptions, lead to similar results, we have what we can call a robust theorem that is relatively free of the details of the model. Hence, our truth is the intersection of independent lies" (Levins 1966; quoted in Weisberg 2013, 157). So far in this chapter, we have considered the epistemic attributes of individual models or alternate aspects of the same model. Robustness is an attribute of a collection of models. The idea is that, taken together, the models increase our confidence in a hypothesis or prediction, because despite their different idealizations and assumptions, all point to the same result.

Philosophers of science have scrutinized robustness in a number of different contexts. Some have presented formalizations of robustness analysis (RA) drawing alternate conclusions. Some are skeptical that RA provides any support for a hypothesis (see e.g. Orzack and Sober 1993), some have tempered skepticism about RA (see e.g. Odenbaugh 2003; Justus 2012), while others conclude that a proper formal treatment of RA shows that RA can provide some confirmatory support (see e.g. Weisberg 2013; Schupbach 2018). Here we spell out the main idea of robustness and briefly examine the role it plays in the assessment of climate models.

Before we turn to model robustness let's set up the general idea of robustness. Jonah Schupbach (2018) does this using the history of experiments on Brownian motion:

> Brownian motion was detected in different fluid media and with different types of small particles, organic and inorganic.

What Makes for a Good (or Bad) Model? **81**

> Brownian motion is robust across various changes to the experimental apparatus (type of particle, medium, container, lighting) and sensitive to others (size of particle, temperature of medium).
>
> *(2018, 277)*

Brownian motion, the haphazard motion of suspended, small particles, was ultimately attributed, by Einstein, to the movement of the molecules in the medium in which the particles were suspended. Schupbach says "To the extent that a scientific result is detected by numerous, diverse means, it is said to be 'robust'" (2018, 276). Schupbach divides cases of robustness into experimental and model driven robustness. In experimental cases, the same result across varying contexts is what we focus on, in modeling he says "the common result is the similar behavior across various models" (2018, 279). When climate scientists say that an ensemble of models, which each include somewhat different assumptions and idealizations, all make the same prediction, the prediction is robust (see Lloyd 2009). Now we will briefly introduce an influential formalism of Levins' notion of model robustness, which has anchored much discussion of model robustness in philosophy of science (cf. Winsberg 2018).

The formulation of model robustness focused on in philosophy of science is due to Orzack and Sober (1993), who developed it in order to make RA precise enough for critical scrutiny. Many philosophers use versions of this formulation (see e.g. Lloyd 2009; J. Odenbaugh 2011; Weisberg 2013; Winsberg 2018): consider a set of models M with members m_i and each of those models contains a common assumption A and at least one distinct assumption. RA says that a prediction, P, or result, R, is robust over M if for each $m_i \in M$, m_i entails P (or R). P (or R) are robust predictions or results and, by Levins' lights, are likely to be true. Lloyd uses this version of RA to argue that climate science results are supported, in part, because they are robust over a large set of climate models (2009). We noted above (Section 5.3) that Lloyd's view of confirmation in Evolutionary Biology is that model confirmation comes from model fit, theoretical assumptions, variety of evidence and robustness. She has a similar view about confirmation in climate science. Here we focus on the role of robustness.

We saw above, in Section 5.3, that Parker criticized Lloyd's approach to model confirmation. Parker (2011) also criticizes Lloyd's

82 What Makes for a Good (or Bad) Model?

claim that robustness confers support on climate results via model agreement. Winsberg summarizes Parker's objection as follows: "the fact that a hypothesis about the climate is supported by a given ensemble of models is not, by itself, sufficient grounds to accept the hypothesis" (2018, 178). Winsberg argues that Parker's objection misses, because rather than defending RA of models alone as providing support for climate science results, Lloyd is arguing that "RA can be part of a story about how a variety of evidence and the agreement of models can provide rational grounds to accept [a] hypothesis" (2018, 178–79). This is consistent with Lloyd's approach to confirmation in evolutionary biology. As we saw above, model fit plays a role in hypothesis confirmation but so does variety of evidence and robustness. In climate science, model robustness is one of the aspects of support for climate hypotheses. Winsberg sides with Lloyd here but thinks that the use of RA can be extended to expand and strengthen Lloyd's account. Winsberg says that the appropriate use of RA in climate science is not just model agreement (as Parker proposes) nor is it to show that a diverse set of models produce the same finding; rather, RA connects all the kinds of diverse evidence for climate hypotheses including modeling predictions and measurements of ocean temperature, artic sea ice, sea level and so on. He proposes an extended, or wide scope, RA from which we can assess the robustness of climate hypotheses across a range of different modes of investigation including modeling.

5.7 Conclusions

Scientists judge the goodness (or badness) of models for many different purposes and philosophers of science propose several alternate approaches to the epistemic appraisal of models. Here we have looked at several philosophical issues arising from the ways in which models are epistemically appraised: accuracy of model description; whether models are confirmed; whether models explain and how well they do so; trade-offs between model desiderata; and whether groups of models are robust. Alternate approaches to epistemic appraisal of models are impacted by many factors, including the type of model up for appraisal, scientists' purposes for their models and methodology in philosophy of science. Some approaches to epistemic appraisal proposed by philosophers of science do not appear to fit scientific practice with respect to epistemic appraisal. We also saw that some model desiderata trade-off, for example, striving for a more general model can trade-off with model precision. Every new model-based scientific discipline, and every new local practice within these disciplines, philosophers turn their attention to will reveal alternate approaches to model appraisal. Future work on model appraisal in philosophy of science will do this, but attention should also be paid to the match, or lack thereof, between philosophical accounts of model appraisal and the approaches used by practicing scientists.

Notes

1 As we noted in Chapter 4, it is hard to provide an account of representation without also invoking epistemic issues.
2 We use the term "pluralistic" here to contrast Lloyd's approach to confirmation with more traditional approaches to confirmation such as Bayesianism or Likelihoodism (see Sober 2008 Chapter 1, for an discussion of alternate analyses of model confirmation). Whereas a Bayesian assesses degrees of confirmation via a probability analysis, Lloyd thinks of confirmation being conferred by fit, which is amenable to Bayesian analysis, but also by robustness and other considerations.
3 This challenge may also be directed at Lloyd's account of model appraisal in Evolutionary Biology but given that Parker and Winsberg's challenge is derived from the details of Climate Science practice, challenging Lloyd's account of model appraisal in Evolutionary Biology should only be made via a detailed look at approaches to model appraisal in Evolutionary Biology.
4 "Skill" is the term Climate Scientists use to characterize their models' appropriateness for prediction and other purposes they use them for.
5 Winsberg's local practice-based approach to epistemic issues in science echoes Alison Wylie's (see e.g. Chapman and Wylie 2016) approach to Archaeology and Melinda Fagan's (see e.g. 2013) approach to Stem Cell Biology.

6

CONCLUSION

Pluralism about Models, Modeling, and Model Evaluation

Science is clearly pursued by modeling. Modeling is certainly not the only way in which scientific knowledge is furthered, but it makes a huge contribution to this task. Focus on models and modeling has enriched philosophy of science. Presenting science as disembodied theories, hypotheses, and observations allowed us to make some headway into epistemological issues in science. However, the assumption that all scientific practice could be accounted for by such a framework was misplaced. Philosophers first turned their attention to models in an attempt to right what they saw as the shortcomings of theory focused, empiricist philosophy of science. However, it was the focus on scientific practice, via attention to modeling, that led to the huge increase in the range of philosophical issues philosophers of science now face. The aim of this book has been to give a glimpse at some of this vast range of philosophical issues.

There is a plurality of models, and in Chapter 2, we provided a small sample of this plurality. This chapter furnished us with examples for discussions in subsequent chapters. Selecting such an eclectic collection of models was also important for shaping discussion in subsequent chapters. If, for example, we had selected only mathematical models in this chapter, we could have concluded that all models are mathematical and perhaps even that all models are some kind of abstract object. The difficulty of the tasks of classifying models and answering the question "What are models?" is much clearer when a wide variety of models are on the table from the outset. It is also important to emphasize the wide range of scientific disciplines that engage in modeling. Modeling is practiced by cognitive scientists, ecologists, economists, neuroscientists, physicists, population geneticists, statisticians, and many more. Modeling is also practiced in many different types of scientific practice. There is highly theoretical modeling, but modeling also plays an important role in experimental science. Modeling can be theoretically informed and completely uniformed by any theoretical assumptions. There are a plurality of models and a plurality of modeling practices.

In Chapter 3, we found little support for adopting a unified account of what models are. Limiting the scope of the term "scientific model" leads to problems. One of which is a demarcation issue. If we say that all scientific models are mathematical models, then we are, implicitly at least, committed to the view that practices that invoke nonmathematical models are not scientific. We also found that there is no adequate classification scheme for models. Classifying models by appealing to their structure or constituents leaves us with cases that fall into two of our proposed classification schemas or reveals very similar models that still do not appear to belong in the same class. Finally, we found that the relation between models and theories is not straightforward and cannot be easily accounted for. For example, the idea that theories are collections of models implies that modeling is part of theorizing, but we know that much modeling takes place in the absence of theoretical considerations. Classifying models via their function or the purposes scientists put them to appears promising but results in an enormous list of model types that is open-ended. This list can always be added to by philosophers who unearth an alternate type of modeling practice. We are faced with a plurality of model types and a plurality of modeling practices in the sciences.

We examined the influential idea that models represent in Chapter 4. The assumption that models represent led to a huge amount of work in philosophy of science on understanding how models represent. We discussed a sample of this work in this chapter. There is some overlap between this work in philosophy of science and other work in philosophy on representation, for example, in aesthetics and in the philosophy of mind. There is no consensus view about model representation.

86 Conclusion

Many philosophers of science hold use or agent-based accounts of representation while others defend similarity-based views. Some hold that models represent via both a similarity relation and the intentions of scientists. We also introduced the consensus-challenging idea that some models do not do their epistemic work via representation. The way in which this challenge is made risks blurring the line between the issue of representation and the appraisal of representation, for example, the appraisal of accuracy.

The issue of model appraisal was tackled in Chapter 5. We began by introducing and discussing accuracy, following from the discussion in Chapter 4. We then considered the extent to which models are confirmed and whether they provide explanations. These discussions revealed close connections between accounts of model appraisal and competing philosophical views on confirmation and explanation respectively. Next, we examined the idea that model desiderata trade-off, for example, a more general model cannot also be a more precise model. Finally, we introduced and discussed model robustness, the idea that a prediction or result can be robust over a collection of models. We saw some have argued that robustness analysis can, and should, play an important role in climate science. There is a plurality of approaches to model appraisal in the sciences. Some of these approaches are dictated by the type of model scientists are working with, but most are dictated by the purposes scientists have for their models. For example, assessing a model used to generate predictions in terms of its explanatory power is likely inappropriate.

Focusing on model-based science has led many philosophers of science to pluralist conclusions. Work on models has revealed a plurality of models, a plurality of model types, a plurality of relations between models and theories, a plurality of accounts of model representation and a plurality of desiderata for model appraisal. Further pluralities are revealed in each of these areas of inquiry. For example, we found a number of alternate accounts of explanation, leading to a number of different ways of distinguishing good from bad explanations.

Winsberg says, "A Martian, visiting earth, who tried to learn about the range of scientific activities in which humans engage by visiting a meeting of the Philosophy of Science Association, would find us to be very parochial in our interests" (2018, 3). The Philosophy of Science

Association meeting is a biennial meeting of philosophers of science held in the USA. Members, both students and faculty, may submit papers on any topic in the philosophy of science. The Philosophy of Science Association was founded in 1933 and held its first biennial meeting in 1968 (Philosophy of Science Association n.d.). For many years, the program at Philosophy of Science Association meetings was dominated by papers in philosophy of physics and papers on inductive and deductive logic. Partly as a result of the shift in focus from the structure and confirmation of theories to models and the practice of model-based scientists, the topics on the meeting program have expanded. Despite this expansion, Winsberg is right, that the meeting program is not a good gauge of the diversity of scientific practice. One way we can expand the scope of work in philosophy of science is by diversifying our scientific interests. Winsberg is comforted by the fact that an increasing number of philosophers are focusing on climate science but there are vast areas of scientific practice still to be explored. There is no good reason to believe that an as yet to be explored area of scientific practice will have no interesting philosophical issues. The practice of any model-based science will provide materials for exploring at least the philosophical issues discussed in this book and doubtless many more.

The agenda for new work in philosophy of science is not simply set by the injunction to explore diverse scientific practices. Research programs in philosophy of science are built upon previous work in philosophy of science. For example, much of the work on model representation we discussed in Chapter 4 responds to earlier work on model representation and representation in general. Philosophy of science is no different from other areas of philosophy in this regard: New research in epistemology builds upon previous work in epistemology. This is in part how all disciplines and subdisciplines set their boundaries. However, just as new cases and examples impact approaches in epistemology, new examples of scientific practice impact work in philosophy of science. Some of the debates about representation we discussed in Chapter 4 have a life of their own, without any reference to science, scientific examples, or scientific practice, whereas others hinge upon examples of scientific practice. Debate over whether representation can be accounted for by similarity is an example of the former and the debate between Cartwright and Morrison over whether

88 Conclusion

theoretical models represent is an example of the latter. Cartwright and Morrison's disagreement is crucially informed by the interpretation of details of the practice in a specific area of physics.

There are myriad opportunities for further work in philosophy of model-based science. These opportunities come from both the expansion of the range of scientific activities we explore and the refinement and extension of established research traditions in philosophy of science. Opportunities also come from the combination of these two approaches. As we saw a number of times in this book, accounts of confirmation, explanation, and so on derived from particular research traditions in philosophy of science are challenged, refined, and sometimes rejected in the face of alternate model-based science practice. In this book we have laid out some of the philosophical terrain that can be explored adopting any of these approaches.

Further Reading

Here are some recommended books from the past and present of model-based philosophy of science. Some of these books focus exclusively on models and model-based science and in others, an examination of model-based science plays a central role.

Cartwright, N. (1983). *How the Laws of Physics Lie.*

Nancy Cartwright's book is one of the first to take both a model-based approach to understanding science and a close look at examples from scientific practice. Cartwright introduces novel accounts of the relations between theories and the world and scientific idealizations along with a fictionalist account of models: models as fables.

Fagan, M. (2013). *Philosophy of Stem Cell Biology: Knowledge in Flesh and Blood.*

Melinda Fagan distinguishes between alternate modeling practices that underlie different fields (and subfields) in biology. She develops an account of models that captures model stem cells appealed to by experimental biologists and contrasts these models with "abstract" models used by theoretical biologists and systems biologists.

Giere, R. (1988). *Explaining Science.*

Ron Giere, influenced by Cartwright's work above, among others, puts scientific practice front and center in his model based approach to understanding science. He defends a similarity-based account of model representation supported by examples from physics as well as geology. He also proposes his approach as a realist alternative to anti-realists in Science Studies, such as Sociologists of Science.

Lloyd, E. A. (1988). *The Structure and Confirmation of Evolutionary Theory.*

Elisabeth Lloyd brings a model-based approach to bear on philosophical issues arising from evolutionary biology. She sheds light on issues such as the units of selection, genotype–phenotype relations, confirmation of evolutionary hypotheses and the explanatory role population genetics can play.

Morgan, M. S. (2012) *The World in the Model.*

Mary Morgan provides both a historical account of the role of various different modeling practices and types of models in economics along with a philosophical analysis of modeling driven by examples taken from economics. Economics is a heavily model-driven practice and provides fertile ground for her examination of issues such as what models are and what their various epistemic roles are.

Morrison, M. (2015). *Reconstructing Reality: Models, Mathematics, and Simulations.*

Margaret Morrison covers a wide range of philosophical issues about modeling in the sciences in this book. One of her primary interests is to develop and defend an alternate account of model representation to that offered by Giere and extended by Weisberg. She also develops an account of how it is that mathematical systems can act as scientific representations.

Odenbaugh, J. (2019). *Ecological Models.*

Jay Odenbaugh explores the role different types of models play in ecology. Ecology is a model-rich area of biology and alternate ecological models provide plenty of material to drive discussion of philosophical issues about modeling. Odenbaugh has particular interest in the role of model robustness in supporting alternate accounts of ecological phenomena.

Potochnik, A. (2017). *Idealization and the Aims of Science.*

Angela Potochnik develops and defends an account of idealization in this book. While her book is not primarily about models, the relevant idealizations that she focuses on in science are present in models. Potochnik's approach is based in an understanding of scientific practice. She emphasizes an approach to philosophy of science that takes the day-to-day work of scientists – cognitively limited beings just like the rest of us – seriously.

Weisberg, M. (2013). *Simulation and Similarity: Using Models to Understand the World.*

Michael Weisberg draws on Giere's model-based view, among others, and adds structure to Giere's similarity-based view of model representation. He also adds to epistemological discussions about models through his treatment of the notion of robustness.

Winsberg, E. (2018). *Philosophy and Climate Science.*

Eric Winsberg does not exclusively focus on climate models in this book, but he does pay plenty of attention to them. Climate science is yet another model-rich field and many of the epistemological problems that climate scientists face arise from the development and appraisal of models. Winsberg provides analyses of many of these issues and also introduces and appraises various other philosophers' contributions to discussions of climate modeling.

REFERENCES

Bailer-Jones, Daniela. 2009. *Scientific Models in Philosophy of Science*. Pittsburgh, PA: University of Pittsburgh Press.

Batterman, Robert W. 2002. "Asymptotics and the Role of Minimal Models." *The British Journal for the Philosophy of Science* 53 (1): 21–38. doi:10.1093/bjps/53.1.21.

Batterman, Robert W., and Collin C. Rice. 2014. "Minimal Model Explanations." *Philosophy of Science* 81 (3): 349–76. doi:10.1086/676677.

Bokulich, Alisa. 2011. "How Scientific Models Can Explain." *Synthese* 180 (1): 33–45. doi:10.1007/s11229-009-9565-1.

———. 2017. "Models and Explanation." In *The Springer Handbook of Model-based Science*, edited by Lorenzo Magnani and Tommaso Bertolotti, 103–18. Dordrecht: Springer.

Boltzmann, L. 1905. "Theories as Representations." In *Philosophy of Science*, edited by A. Danto and S. Morgenbesser, 245–52. New York: Meridian Books.

———. 1911. "Model." In *Encyclopedia Britannica*, 638–40.

Box, George E. P., and Norman R Draper. 1986. *Empirical Model-Building and Response Surface*. New York: John Wiley & Sons, Inc.

Callender, Craig, and Jonathan Cohen. 2006. "There Is No Special Problem about Scientific Representation." *THEORIA. An International Journal for Theory, History and Foundations of Science* 21 (1): 67–85. doi:10.1387/theoria.554.

Campbell, Norman Robert. 1920. *Physics: The Elements*. 1st edition. Cambridge: Cambridge University Press.

Cartwright, Nancy. 1983. *How the Laws of Physics Lie*. Oxford: Clarendon Press.

92 References

———. 1999. "Models and the Limits of Theory: Quantum Hamiltonians and the BCS Models of Superconductivity." In *Models as Mediators: Perspectives on Natural and Social Science*, edited by Mary S Morgan and Margaret Morrison, 241–81. Cambridge: Cambridge University Press.

Chapman, Robert, and Alison Wylie. 2016. *Evidential Reasoning in Archaeology*. Debates in Archaeology. London; New York: Bloomsbury Academic, an imprint of Bloomsbury Publishing Plc.

Costa, Newton C. A. da, and Steven French. 2003. *Science and Partial Truth: A Unitary Approach to Models and Scientific Reasoning*. Oxford Studies in Philosophy of Science. Oxford; New York: Oxford University Press.

Cotman, Carl W., and James L. McGaugh. 1980. *Behavioral Neuroscience: An Introduction*. New York: Academic Press.

Craver, Carl F. 2009. *Explaining the Brain*. Oxford: Oxford University Press.

Craver, Carl F. and Tabery, James 2019. "Mechanisms in Science." In *The Stanford Encyclopedia of Philosophy*, edited by Edward N. Zalta, Summer 201. Metaphysics Research Lab, Stanford University. https://plato.stanford.edu/archives/sum2018/entries/mechanisms-science/.

Cummins, R. 1989. *Meaning and Mental Representation*. Cambridge, MA: MIT Press.

Darden, Lindley. 2002. "Strategies for Discovering Mechanisms: Schema Instantiation, Modular Subassembly, Forward/Backward Chaining." *Philosophy of Science* 69 (S3): S354–65. doi:10.1086/341858.

Doncaster, C. Patrick, and Andrew J. H Davey. 2007. *Analysis of Variance and Covariance: How to Choose and Construct Models for the Life Sciences*. Cambridge: Cambridge University Press.

Downes, Stephen M. 1992. "The Importance of Models in Theorizing: A Deflationary Semantic View." *PSA: Proceedings of the Biennial Meeting of the Philosophy of Science Association* 1992 (1): 142–53.

———. 2009. "Models, Pictures, and Unified Accounts of Representation: Lessons from Aesthetics for Philosophy of Science." *Perspectives on Science* 17 (4): 417–28. doi:10.1162/posc.2009.17.4.417.

Fagan, Melinda Bonnie. 2013. *Philosophy of Stem Cell Biology: Knowledge in Flesh and Blood*. Houndmills: Palgrave Macmillan.

———. 2016. "Generative models: Human embryonic stem cells and multiple modeling relations." *Studies in History and Philosophy of Science Part A* 56: 122–134. doi: 10.1016/j.shpsa.2015.10.003

Fine, Arthur. 1993. "Fictionalism." *Midwest Studies in Philosophy* 18 (1): 1–18. doi:10.1111/j.1475-4975.1993.tb00254.x.

———. 2009. "Science Fictions: Comment on Godfrey-Smith." *Philosophical Studies* 143: 117–25.

Forber, Patrick. 2010. "Confirmation and Explaining How Possible." *Studies in History and Philosophy of Science Part C: Studies in History and*

References **93**

Philosophy of Biological and Biomedical Sciences 41 (1): 32–40. doi:10.1016/j.shpsc.2009.12.006.

French, S., and J. Ladyman. 1999. "Reinflating the Semantic Approach." *International Studies in Philosophy of Science* 13 (2): 103–22.

Frigg, Roman. 2009. "Models and Fiction." *Synthese* 172 (2): 251. doi:10.1007/s11229-009-9505-0.

Frigg, Roman, and James Nguyen. 2017. "Models and Representation." In *Springer Handbook of Model-Based Science*, edited by Lorenzo Magnani and Tommaso Bertolotti, 49–102. Dordrecht: Springer.

Frigg, Roman, and Stephan Hartmann. 2018. "Models in Science." In *The Stanford Encyclopedia of Philosophy*, edited by Edward N. Zalta, Summer 2018. Metaphysics Research Lab, Stanford University. https://plato.stanford.edu/archives/sum2018/entries/models-science/.

Futuyma, Douglas J. 1998. *Evolutionary Biology.* Sunderland, MA: Sinauer.

———. 2006. *Evolutionary Biology.* New York; Basingstoke: W.H. Freeman.

Giere, R. N. 1988. *Explaining Science.* Chicago, IL: University of Chicago Press.

———. 1999. "Using Models to Represent Reality." In *Model-Based Reasoning in Scientific Discovery*, edited by L. Magnaiv, N. Nersessian, and P. Thagard, 41–57. New York: Plenum Publishers.

———. 2004. "How Models Are Used to Represent Reality." *Philosophy of Science* 71: 742–51.

———. 2009. "An Agent-Based Conception of Models and Scientific Representation." *Synthese* 172 (2): 269. doi:10.1007/s11229-009-9506-z.

Giere, Ronald N, John Bickle, and Robert F Mauldin. 2006. *Understanding Scientific Reasoning.* Southbank: Thomson/Wadsworth.

Glennan, Stuart. 2005, January. "Modeling Mechanisms." *Studies in History and Philosophy of Biological and Biomedical Sciences*, 443–64. doi:10.1016/j.shpsc.2005.03.011.

Godfrey-Smith, P. 2006. "The Strategy of Model-Based Science." *Biology and Philosophy* 21: 725–40.

Goodman, N. 1976. *Languages of Art.* Indianapolis, IN: Hackett.

Grüne-Yanoff, Till. 2013. "Appraising Models Nonrepresentationally." *Philosophy of Science* 80 (5): 850–61. doi:10.1086/673893.

Hacking, Ian. 1983. *Representing and Intervening: Introductory Topics in the Philosophy of Natural Science.* Cambridge: Cambridge University Press.

Hempel, C. G. 1966. *Philosophy of Natural Science.* Engelwood Cliffs, NJ: Prentice Hall.

Hesse, Mary B. 1966. *Models and Analogies in Science.* Notre Dame, IN: University of Notre Dame Press.

Hughes, R. I. G. 1997. "Models and Representation." *Philosophy of Science* 64: S325–36.

94 References

Hurford, Amy. 2012. "Overview of Mathematical Modelling in Biology II." *Just Simple Enough: The Art of Mathematical Modelling* (blog). https://theartofmodelling.wordpress.com/2012/01/04/overview-of-mathematical-modelling-in-biology-ii/.

Justus, James. 2012. "The Elusive Basis of Inferential Robustness." *Philosophy of Science* 79 (5): 795–807. doi:10.1086/667902.

Kahneman, Daniel. 1973. *Attention and Effort*. Prentice-Hall Series in Experimental Psychology. Englewood Cliffs, NJ: Prentice-Hall.

Kaplan, J. 2008. "The End of the Adaptive Landscape Metaphor." *Biology and Philosophy* 23: 625–38.

Keller, Evelyn Fox. 2000. "Models of and Models for: Theory and Practice in Contemporary Biology." *Philosophy of Science* 67: S72–86.

Kitcher, Philip. 1993. *The Advancement of Science: Science without Legend, Objectivity without Illusions*. New York; Oxford: Oxford University Press.

Ladyman, James. 1998. "What Is Structural Realism?" *Studies in History and Philosophy of Science Part A* 29 (3): 409–24.

———. 2007. "Structural Realism," November. https://stanford.library.sydney.edu.au/entries/structural-realism/.

Leonelli, Sabina, and Rachel A. Ankeny. 2013. "What Makes a Model Organism?" *Endeavour* 37 (4): 209–12. doi:10.1016/j.endeavour.2013.06.001.

Levins, Richard. 1966. "The Strategy of Model Building in Population Biology." *American Scientist* 54 (4): 421–31.

Lewontin, R. C. 1963. "Models, Mathematics and Metaphors." *Synthese* 15 (2): 222–44.

Lloyd, E. A. 1988. *The Structure and Confirmation of Evolutionary Theory*. Princeton, NJ: Princeton University Press.

———. 2009. "Varieties of Support and Confirmation of Climate Models." *Proceedings of the Aristotelian Society Supplementary Volume* 83: 217–36.

———. 2010. "Confirmation and Robustness of Climate Models." *Philosophy of Science* 77: 971–84.

———. 2017. "Models in the Biological Sciences." In *The Springer Handbook of Model-based Science*, edited by Lorenzo Magnani and Tommaso Bertolotti, 913–28. Dordrecht: Springer.

Lynch, Michael and Walsh, Bruce 1998. *Genetics and Analysis of Quantitative Traits*. Sunderland, MA: Sinauer.

Machamer, Peter, Lindley Darden, and Carl F. Craver. 2000. "Thinking about Mechanisms." *Philosophy of Science* 67 (1): 1–25. doi:10.1086/392759.

MacLeod, Miles Alexander James. 2018. "Scientific Subordination, Molecular Biology and Systems Biology." In *Scientific Imperialism: Exploring the Boundaries of Interdisciplinarity*, 187–204. https://research.utwente.nl/en/publications/scientific-subordination-molecular-biology-and-systems-biology.

Matthewson, J., and M. Weisberg. 2009. "The Structure of Tradeoffs in Model Building." *Synthese* 170: 169–90.

References 95

McGreevy, Joe W., Chady H. Hakim, Mark A. McIntosh, and Dongsheng Duan. 2015. "Animal Models of Duchenne Muscular Dystrophy: From Basic Mechanisms to Gene Therapy." *Disease Models & Mechanisms* 8 (3): 195–213. doi:10.1242/dmm.018424.

McKinley, M. J., and A. K. Johnson. 2004. "The Physiological Regulation of Thirst and Fluid Intake." *News in Physiological Sciences* 19 (1): 1–6. doi:10.1152/nips.01470.2003.

McMullin, Ernan. 1985. "Galilean Idealization." *Studies in History and Philosophy of Science Part A* 16 (3): 247–73. doi:10.1016/0039-3681(85)90003-2.

Mitchell, Sandra D. 2013. *Unsimple Truths: Science, Complexity, and Policy.* Chicago, IL: University of Chicago Press.

Morgan, Mary S. 2012. *The World in the Model: How Economics Work and Think.* Cambridge: Cambridge University Press.

Morgan, Mary S., and M. Morrison. 1999. "Models as Mediating Instruments." In *Models as Mediators*, edited by M. S. Morgan and M. Morrison, 10–37. Cambridge: Cambridge University Press.

Morrison, Margaret. 2015. *Reconstructing Reality: Models, Mathematics, and Simulations.* Oxford: Oxford University Press.

Morrison, Margaret, and M. Morgan Eds. 1999a. *Models as Mediators.* Cambridge: Cambridge University Press.

Morrison, Margaret, and Mary S. Morgan. 1999b. "Introduction." In *Models as Mediators*, edited by M.S. Morgan and M. Morrison, 1–10. Cambridge: Cambridge University Press.

O'Connor, Cailin, and James Owen Weatherall. 2016. "Black Holes, Black-Scholes, and Prairie Voles: An Essay Review of Simulation and Similarity, by Michael Weisberg." *Philosophy of Science* 83 (4): 613–26. doi:10.1086/687265.

Odenbaugh, Jay. 2003. "Complex Systems, Trade-Offs and Mathematical Modeling: A Response to Sober and Orzack." *Philosophy of Science* 70: 1496–1507.

———. 2011. "True Lies: Realism, Robustness, and Models." *Philosophy of Science* 78 (5): 1177–88. doi:10.1086/662281.

———. 2018. "Models, Models, Models: A Deflationary View." *Synthese*, 1–16. doi:10.1007/s11229-017-1665-8.

———. 2019. *Ecological Models.* Cambridge University Press.

Orzack, Steven Hecht, and Elliott Sober. 1993. "A Critical Assessment of Levins's the Strategy of Model Building in Population Biology (1966)." *The Quarterly Review of Biology* 68 (4): 533–46.

Otto, S. P., and T. Day. 2007. *A Biologists's Guide to Mathematical Modeling in Ecology and Evolution.* Princeton, NJ: Princeton University Press.

Parker, Wendy S. 2009. "Confirmation and Adequacy-for-Purpose in Climate Modelling." *Proceedings of the Aristotelian Society Supplementary Volume* 83: 233–49.

96 References

———. 2011. "When Climate Models Agree: The Significance of Robust Model Predictions★." *Philosophy of Science* 78 (4): 579–600. doi:10.1086/661566.

———. 2015. "Getting (Even More) Serious about Similarity." *Biology & Philosophy* 30 (2): 267–76. doi:10.1007/s10539-013-9406-y.

Perini, L. 2005. "The Truth in Pictures." *Philosophy of Science* 71: 262–85.

Peschard, Isabelle. 2011. "Making Sense of Modeling: Beyond Representation." *European Journal for Philosophy of Science* 1 (3): 335–52. doi:10.1007/s13194-011-0032-8.

Philosophy of Science Association. n.d. "Philosophy of Science Association." Accessed April 7, 2019. https://www.philsci.org/.

Potochnik, Angela. 2012. "Feminist Implications of Model-Based Science." *Studies in History and Philosophy of Science Part A* 43 (2): 383–89. doi:10.1016/j.shpsa.2011.12.033.

———. 2017. *Idealization and the Aims of Science*. Chicago, IL: The University of Chicago Press.

Ratti, Emanuele. 2018. "'Models of' and 'Models for': On the Relation between Mechanistic Models and Experimental Strategies in Molecular Biology." *The British Journal for the Philosophy of Science*. doi:10.1093/bjps/axy018.

Redhead, Michael. 1980. "Models in Physics." *British Journal for Philosophy of Science* 31: 145–63. doi:10.1093/bjps/31.2.145.

Rosenblueth, Arturo, and Norbert Wiener. 1945. "The Role of Models in Science." *Philosophy of Science* 12 (4): 316–21. doi:10.1086/286874.

Schupbach, Jonah N. 2018. "Robustness Analysis as Explanatory Reasoning." *The British Journal for the Philosophy of Science*, 275–300. doi:10.1093/bjps/axw008.

Seger, J., and J. W. Stubblefield. 1996. "Optimization and Adaptation." In *Adaptation*, edited by M. R. Rose and G. V. Lauder, 93–123. New York: Academic Press.

Slater, P. J. B. 2004. *Essentials of Animal Behaviour*. Cambridge: Cambridge University Press.

Smith, J. Maynard, and G. R. Price. 1973. "The Logic of Animal Conflict." *Nature* 246 (5427): 15–18. doi:10.1038/246015a0.

Smith, John Maynard, and G. A. Parker. 1976. "The Logic of Asymmetric Contests." *Animal Behaviour* 24 (1): 159–75. doi:10.1016/S0003-3472(76)80110-8.

Sober, Elliott. 2008. *Evidence and Evolution: The Logic Behind the Science*. Cambridge; New York: Cambridge University Press.

Solomon, Miriam. 2001. *Social Empiricism*. Cambridge, MA: MIT Press. http://search.ebscohost.com/login.aspx?direct=true&scope=site&db=nlebk&db=nlabk&AN=75049.

Steele, Katie, and Charlotte Werndl. 2016. "The Diversity of Model Tuning Practices in Climate Science." *Philosophy of Science* 83 (5): 1133–44.

References **97**

Sterrett, S. G. 2017. "Physically Similar Systems – A History of the Concept." In *The Springer Handbook of Model-based Science*, edited by Lorenzo Magnani and Tommaso Bertolotti, 377–412. Dordrecht: Springer.

Strayer, David L., and Frank A. Drews. 2008. "Attention." In *Handbook of Applied Cognition*, 29–54. Wiley-Blackwell. doi:10.1002/9780470 713181.ch2.

Strevens, Michael. 2008. *Depth: An Account of Scientific Explanation*. Cambridge, MA: Harvard University Press.

Suárez, M. 2003. "Scientific Representation: Against Similarity and Isomorphism." *International Studies in Philosophy of Science* 17: 225–44.

Suárez, Mauricio, and Nancy Cartwright. 2008. "Theories: Tools versus Models." *Studies in History and Philosophy of Science Part B: Studies in History and Philosophy of Modern Physics* 39 (1): 62–81. doi:10.1016/j.shpsb.2007.05.004.

Suppe, F. 1977a. "The Search for Philosophic Unverstanding of Scientific Theories." In *The Structure of Scientific Theories*, edited by F. Suppe, 2–232. Urbana: University of Illinois Press.

———. 1977b. *The Structure of Scientific Theories*. 2nd edition. Urbana: University of Illinois Press.

Teller, P. 2001. "Twilight of the Perfect Model." *Erkenntnis* 55: 393–415.

Thomson-Jones, Martin. 2010. "Missing Systems and the Face Value Practice." *Synthese* 172 (2): 283–99.

Toon, Adam. 2012. *Models as Make-Believe*. London: Palgrave Macmillan.

Tversky, Amos. 1977. "Features of Similarity." *Psychological Review* 84 (4): 327–52. doi:10.1037/0033-295X.84.4.327.

Van Fraassen, Bas C. 1980. *The Scientific Image*. Oxford: Oxford University Press.

Van Fraassen, B. 2008. *Scientific Representation*. Oxford: Oxford University Press.

Watson, J. D., and F. H. C. Crick. 1953. "Molecular Structure of Nucleic Acids: A Structure for Deoxyribose Nucleic Acid." *Nature* 171 (4356): 737–38. doi:10.1038/171737a0.

Weisberg, Michael. 2007. "Who Is a Modeler?" *British Journal for Philosophy of Science* 58: 207–33.

———. 2012. "Getting Serious about Similarity." *Philosophy of Science* 79 (5): 785–94. doi:10.1086/667845.

———. 2013. *Simulation and Similarity: Using Models to Understand the World*. Oxford Studies in Philosophy of Science. Oxford: Oxford University Press.

———. 2015. "Biology and Philosophy Symposium on Simulation and Similarity: Using Models to Understand the World: Response to Critics." *Biology & Philosophy* 30 (2): 299–310. doi:10.1007/s10539-015-9475-1.

Weisberg, Michael, and Kenneth Reisman. 2008. "The Robust Volterra Principle." *Philosophy of Science Philosophy of Science* 75 (1): 106–31.

Wimsatt, W. C. 2007. "False Models as Means to Truer Theories." In *Re-Engineering Philosophy for Limited Beings*, edited by W. C. Wimsatt, 94–132. Cambridge, MA: Havard University Press.

Winsberg, Eric. 2010. *Science in the Age of Computer Simulation*. Chicago, IL: University of Chicago Press.

———. 2018. *Philosophy and Climate Science*. Cambridge: Cambridge University Press.

Winther, Rasmus Grønfeldt. 2006. "Parts and Theories in Compositional Biology." *Biology and Philosophy* 21 (4): 471–99.

———. 2016. "The Structure of Scientific Theories." In *The Stanford Encyclopedia of Philosophy*, edited by Edward N. Zalta. Metaphysics Research Lab, Stanford University. https://plato.stanford.edu/archives/win2016/entries/structure-scientific-theories/.

INDEX

Note: *Italic* page numbers refer to figures and page numbers followed by "n" denote endnotes.

abstract direct representation (ADR) 47
accidental representation 57
accounts of model representation 59–63
accurate description 70–2
actual generality (a-generality) 79
adenine, guanine, cytosine, and thymine (A, G, C, T) 6
agent-based accounts 4
agent-based computational models 38
agent-based view of representation 59–63, 86
American Naturalist (journal) 37
analysis of variance (ANOVA) models 24, 41
animal behavior 17, 33
ANOVA model *see* analysis of variance (ANOVA) models
appraising models 4, 66, 71, 73
Army Corps of Engineers 26
array of models 41, 42
atom, Bohr model of 21, 23
attention model 27, *27*, 35, 44
axiomatic system 43

Bailer-Jones, Daniela 36, 50n1
Batterman, Robert W. 77
Bayesian analysis 83n2
biology and economics approaches 17
"biting the bullet" 36–7
Bohr model of atom 21, 23
Bokulich, Alisa 75–7
Brownian motion 80–1
"building blocks" of DNA 6

calibrated/uncalibrated model 71, 72
Callender, Craig 55, 56, 62, 63, 68n8
Campbell, N.R. 53
capacity model of attention *see* Kahneman model
carbon dioxide, concentration of 26
cartoons representations 54
Cartwright, Nancy 43, 50n2, 51n11, 64, 67, 87, 88
ceteris paribus 79
classification schemes for models 3, 35, 40, 49, 85
climate models 24–6, 73–4, 81
coding nucleotide triplet 14
Cohen, Jonathan 55, 56, 62, 63, 68

100 Index

commodity and supply price model 15–16
competition models 12
computer models 36, 39, 40
concrete models 39
confirmation and fit model 73–5
confirmation of laws 42
Copernicus, Nicolaus 21, 35
Cotman, Carl W. 29
COVID 19 32n1
Craver, Carl F. 64, 70, 71, 75–9
Crick, Francis 6
CRISPR-Cas systems 65, 79
crosses and squares represent data 11
Cummins, Robert 68n4

damped oscillator model 21, 46
Darden, Lindley 29
Darwin, Charles 36
Day, Troy 37
deductive logic 35
deflationism about models 47
dehydration 29
demarcation criterion 36
diagrammatic models 3
diploid population growth *11*, 11
DNA methylation 72
DNA model 40, 47, 66, 71; "building blocks" of 6, *8*; double helical structure 6; population growth models 6–8; stick and bead model 6; structure of 6; Watson and Crick model of 6, 7
DNA–RNA process *9*
Duchenne muscular dystrophy (DMD) 29, 30
dynamical system 61

Ecological Models (Odenbaugh) 90
Ecology (journal) 37
Edgeworth–Bowley box 15, *16*, 72, 78
Einstein, Albert 81
electron movements 21
elementary quantum mechanics 23
elliptical motion of planets 21
"empiricists structuralism" 37

epistemic representation (ER) schema 56–7
epistemic virtue of models 56
equation-based model 61
evaluating curve fitting 73
Evolution (journal) 37
evolutionary biology 81–2, 83n3
evolutionary game theoretic models 17–18
evolutionary theory 44
experiment continuum theory 64
Explaining Science (Giere) 88
explanation, goodness of 75–8
explicitly hydraulic models 28
exponential growth model 6; differential equation in 7; of global population *10*; theoretical population, curve for *9*

Fagan, Melinda 49, 83n5
"false" models, treatment of 63
Fine, Arthur 51n11
food-seeking behavior 29
formal model 34, 35; definition of 35
fossil fuel 26
Fox Keller, Evelyn 64–6, 79
Fraschini, F. 30
French, Steven 36, 37
Frigg, Roman 42, 51n9, 51n11, 56, 57, 60, 62, 67
Futuyma, Douglas 35, 36, 38

galilean idealization 76, 77
game theoretic model 17, 33, 76
Giere, Ron 41, 43–5, 47, 50n6, 51n10, 53, 55, 57, 58, 60, 64
Global Climate models 36
global warming 26
Godfrey-Smith, Peter 36, 45, 46, 47, 48, 51n10
Goodman, Nelson 33, 57
graphical model 35, 36
greenhousing process 25, *25*
Grüne-Yanoff, Till 64, 66

Hacking, Ian 43
haploid population growth *11*, 11

Index **101**

"harmless" falsehoods 77
harmonic oscillator equation 21, 36
harmonic oscillator model 40, 64
Hartmann, Stephan 42, 51n9, 51n12
Hawks and Doves model 18–19
heat, global absorption and
    radiation of 25
Heliocentric model 22
Hesse, Mary 42, 53
high-fidelity model 71
Hooke's Law 20, *20*
How the Laws of Physics Lie
    (Cartwright) 88
Hughes, R.I.G. 53, 55
Hurford, Amy 69

Idealization and the Aims of Science
    (Potochnik) 89
"imagined objects of literary
    fiction" 48
individual models 80
interpretive models 64
isomorphism, versions of 56

Jenkin, Fleeming 17, *18*
Johnson, A. K. 29, 30

Kahneman, Daniel 27, 29
Kahneman model 27–8
Kaplan, Jonathan 50n4
Kepler, Johannes 21
Kitcher, Philip 36

Ladyman, James 36, 37
large-scale biocide 12
Levins, Richard 78, 80, 81
Lewontin, Richard 35
linear harmonic oscillators 20, 21, 44
Lloyd, Elisabeth 70, 73, 74, 81–2,
    83n1
loci in genomes 14
logistic growth model 6, 40;
    applications of 11; differential
    equation in 8; diploid population
    growth *11*; haploid population
    growth *11*; Lotka-Volterra
    model 12; representations 61; in

theoretical population *10*; yeast,
    example of 11
Lorenz, Konrad 28, 29, 35, 44
Lotka-Volterra model 12, 36, 40,
    73, 78

McGaugh, James L. 29
McKinley, M. J. 29, *31*
McMullin, Ernan 76
maize plants 24
Mangoldt's models 15
"manipulating biological entities" 65
Martini, L. 30
mass and spring system *20*, 20–1
matching approach 71
material models 34, 35
mathematical models 2–3, 34–7, *38*,
    61; agent-based computational
    models 38; Lotka-Volterra model
    40; mathematical structures 39;
    San Francisco Bay model 39; *vs.*
    organismal models *39*
"mathematical problem" 55
Matthewson, John 78, 79
Maynard Smith, John 17, 33
mechanical models 29, 36, 38, 50n1
mechanism schemas 29
mechanisms indicates care 29
mechanistic model 71
mental representations 62
minimalist idealization 76
model-based science 52–3
model generality 78–9
model organisms 40
models: broad scope notion of
    33–4; classification of 45–6, 85;
    formal model 34; Futuyma's
    classification of 36; Goodman's
    description of 33; material models
    34; mathematical description
    33; mathematical models 34;
    nature and function of 33;
    representational inaccuracy 54;
    robustness, notion of 80–2; and
    theories 42–5; types of 34–42;
    Weisberg's classification 38
models-as-representations view 53

102 Index

"models-for" intervention 64, 79
"models-of" phenomena 64
model world relationship 58, 59, 61, 62
molecular biology models 3
molecular models 12
monistic approach 5
Morgan, Mary 42, 48, 50n6, 53
Morrison, Margaret 42, 48, 50n6, 53, 54, 56, 59–63, 67, 68n6, 87, 88
motivation model 35, 44
motivation, "psychohydraulic" model of 28, *28*
Motta, M. 30
multifactor anova models 24
multiple-models idealization 76
multiprogramming modeling environment 19
muscular dystrophy mouse 49
mutation (nucleotide substitution) 14, 15

"naturalistic" accounts of representation 60
natural system 70, 73
neutral evolution models *14*, 14
Neutral Models of Evolution 44
Newtonian model 44, 78
Newton's Second Law 20
Nguyen, James 56, 57, 60, 62, 67
non-coding triplet 14
nonmathematical models 36
nonnatural representations 62
nonrepresentationalist approaches 4
nonrepresentational models 63–5, 67
nucleotides, strands of 6
nucleotide substitutions 14
null model 24

O'Connor, Cailin 41
Odenbaugh, Jay 47
one-factor/one-way ANOVA models 24
organismal models *39*
Orzack, Steven Hecht 78, 81
Otto, Sarah 37

Parker, Wendy 62, 68n5, 73, 74, 77, 81, 82, 83n3
Perini, Laura 68n2
Peschard, Isabelle 64, 65, 71
"philosophical theory of representation" 59
Philosophy and Climate Science (Winsberg) 89
philosophy of science 63, 75, 78, 84
Philosophy of Science Association 87
Philosophy of Stem Cell Biology: Knowledge in Flesh and Blood (Fagan) 89
physical model 35, 36, 53
planetary motion 21; Newtonian model of 78
planets, positions of 21
pluralistic approach 73, 83n1
population growth models 6–8, 61; carrying capacity of 8; competition models 12; DNA–RNA process *9*; modeling behavior in 18
possible generality (p-generality) 79
practice-based approach 3, 50, 83n5
pragmatic approach 61
predator–prey systems 78
prey population 12
probability theory 2
psychohydraulic model of motivation 28, *28*, 28–9

quantum mechanics 21

racial segregation model 19
racism hypothesis for segregation 66
Ratti, Emanuele 64, 65, 70, 71, 78
real systems, descriptions of 70
"received" view of model explanation 77
Reconstructing Reality: Models, Mathematics, and Simulations (Morrison) 89
Redhead, Michael 50n3
representations 52–6; accurate description 70–2; agent-based and use accounts of 59–63;

alternate ER schema 60; cartoons representations 54; epistemic representation 56; modeling without 63–7; "naturalistic" accounts of 60; scientific representations 54, 62; similarity-based views of 55–9; view of models 4

"requirement of directionality" 57

Rice, Collin C. 77

RNA molecules 6

robustness analysis (RA) 80, 81

robustness, notion of 80–2

robust theorem 80

Rosenblueth, Arturo 34, 35

Rutherford's model of the atom 21, *23*

San Francisco Bay model *26*, 26–7, 30, 39, 44, 70, 71

Schelling model of segregation 19, *19*, 40, 66

Schupbach, Jonah N. 80, 81

"scientific model" 33, 41; classification schemes 3; logical empiricist philosophy of 42; mathematical models 2–3; molecular biology models 3; philosophers of 1–2, 44; philosophical issues in 2; probability theory 2; representations 54, 56; sentence-based understanding of 1

scientific theories 2; axiomatic system 43; definition of 45; 'model-building too kits' 43; "received view" of 42–3; semantic view of 43; structure of 43

semantic view of scientific theories 43

sentence-based approach 1–2

similarity-based views of representation 55–6, 86

Simulation and Similarity: Using Models to Understand the World (Weisberg) 89

Smith, Adam 15

Sober, Elliott 41, 50n4, 78, 81

solar radiation 25

solar system model 21, 31, 78; Newton's model of 21

statistical models 24, 41, 73

Sterrett, Susan 68n3

stick and bead model 6

"stipulative fiat view" 62, 63

Strevens, Michael 76, 77, 79

The Structure and Confirmation of Evolutionary Theory (Lloyd) 88

studying model-based science 74

Suarez, Mauricio 57, 60

supply and demand model 15–17, *18*

target system, descriptions of 70

Teller, Paul 43, 50n8, 52–3, 59, 60

testing ANOVA models 24

theoretical population: exponential growth curve for *9*; logistic growth curve in *10*

"theory of representation seems unnecessary" 59

thirst model 29, 31

three-way distinction 41, 42

"tools for material change" 64

trade-offs, model desiderata 78–80; ecology and population biology 78; precision and generality 79

Tversky, Amos 35, 44, 58

two-body system 21

two-factor ANOVA model 24

Van Fraassen, Bas 37, 54, 56, 59–63, 68n1

Velázquez, Diego 57

verbal models 35, 47

Watson and Crick model 6, 7, 12, 27

Watson, James 6, 7, 12

way models 55

Weatherall, James 41

Weiner, Norbert 34, 35, 50n1

Weisberg, Michael 12, 13, 19, 38–42, 45, 49, 55, 58–60, 62, 68n5, 71, 72, 76, 78, 79; abstract direct representation 47; calibrated and uncalibrated modeling 71; classification scheme 40, 41; computational models 39–41; concrete models 39; epistemic focus 40; Lotka-Volterra model 12; mathematical models 39–41; model generality 78, 79; model organisms 40; model world relationship 58, 62; representational accuracy 58; set theory 58; structures and properties 40; three-way distinction 41, 42; verbal models 47

Wimsatt, William 63, 66, 69, 72, 77

Winther, Rasmus Grønfeldt 47, 50n7

"work of fiction" 72

The World in the Model (Morgan) 89

wrong/false model 72

Wylie, Alison 83n5